Volatile Organic Compounds in the

ISSUES IN ENVIRONMENTAL SCIENCE AND TECHNOLOGY

How to obtain future titles on publication

A subscription is available for this series. This will bring delivery of each new volume immediately upon publication. For further information, please write to:

The Royal Society of Chemistry
Turpin Distribution Services Limited
Blackhorse Road
Letchworth
Herts SG6 1HN, UK

Telephone: +44 (0) 1462 672555
Fax: +44 (0) 1462 480947

ISSUES IN ENVIRONMENTAL SCIENCE AND TECHNOLOGY

EDITORS: R. E. HESTER AND R. M. HARRISON

4

Volatile Organic Compounds in the Atmosphere

THE ROYAL
SOCIETY OF
CHEMISTRY

ISBN 0-85404-215-6
ISSN 1350-7583

A catalogue record for this book is available from the British Library

Published by The Royal Society of Chemistry, Thomas Graham House,
Science Park, Milton Road, Cambridge CB4 4WF, UK

Typeset in Great Britain by Vision Typesetting, Manchester
Printed and bound in Great Britain by Bath Press, Bath

Preface

Whilst volatile organic compounds (VOCs) have never had the high profile of some other pollutants which have attracted attention from pressure groups and the media, collectively they represent one of the most important groups of trace atmospheric constituents. They are important in all parts of the globe and over a wide range of altitudes. Some are appreciably toxic in their own right and the UK Expert Panel on Air Quality Standards has recommended guidelines for benzene and 1,3-butadiene in the atmosphere which have been accepted by the British government; some other countries have also set air quality standards for benzene. Other VOCs are important primarily because of their atmospheric reactivity and consequent influence on the concentrations of tropospheric photochemical ozone, both in pollution episodes and in the background atmosphere. The photochemical ozone creation potential concept seeks to quantify this influence. Moving to higher altitudes, the impact of chlorofluorocarbons (CFCs) on stratospheric ozone has been a crucial one and, thanks to the Montreal Protocol, the ozone layer should be protected from this influence. However, CFCs play an important role both as industrial chemicals and within consumer products and it has proved difficult to find replacements which offer the same benefits of non-inflammability, high stability, and low toxicity, but which have a benign influence on the atmosphere.

Within this Issue we seek to explore many of the scientific aspects relating to volatile organic compounds in the atmosphere. In the first article, Dick Derwent of the Meteorological Office provides a broadly-based introduction to the atmospheric cycle of VOCs by considering their sources, distribution, and fates. This sets the scene for more specialized subsequent articles. Recent years have seen a growing appreciation of the importance of naturally generated VOCs in the atmosphere. In some areas with warm climates, VOCs from vegetation can play an equal or greater role than anthropogenic sources in contributing to low-level ozone formation. Much still needs to be learned about the chemistry and fluxes of these natural VOCs and Nick Hewitt and Xu-Liang Cao of Lancaster University provide a state of the art review of current knowledge. The recent availability of automated instrumentation for monitoring VOCs in urban air has led to a rapid expansion of our database and knowledge. Geoff Dollard and colleagues from the National Environmental Technology Centre explain the UK hydrocarbon monitoring network, one of the most advanced networks in the world, and discuss some of the early data from it.

Many countries are now signatories to international agreements to limit and ultimately reduce emissions of VOCs to the atmosphere. Such controls can only be effective in the context of high quality source inventory information and Neil Passant of the National Environmental Technology Centre reviews information on source inventories and their development and considers control strategy options for VOCs. The atmospheric chemistry of VOCs is crucial to a full appreciation of their behaviour and two articles deal, respectively, with the tropospheric and stratospheric behaviour of important VOC compounds. Roger Atkinson of the University of California reviews the gas phase tropospheric chemistry of organic compounds whilst Pauline Midgley, an independent consultant, considers the impact of CFCs and their alternatives on the chemistry and physics of the stratosphere and troposphere.

Recent work has shown that construction materials and furnishings can act as a major source of VOCs in indoor air, and concentrations of some compounds indoors may greatly exceed outdoor concentrations. Derrick Crump of the Building Research Establishment has led a major programme of research on this topic and presents data from his and other studies in an article on VOCs in indoor air. In the final article, John Murlis describes the policy implications of VOCs and the development of policy in the UK.

We believe that this Issue has assembled some of the most up-to-date and relevant material from the large body of information now currently available on atmospheric VOCs. Each of the authors is a recognized expert in his or her particular area and we feel confident that this Issue will prove extremely valuable to our widely-based readership.

Ronald E. Hester
Roy M. Harrison

Contents

Contents

Editors

Ronald E. Hester, BSc, DSc(London), PhD(Cornell), FRSC, CChem

Ronald E. Hester is Professor of Chemistry in the University of York. He was for short periods a research fellow in Cambridge and an assistant professor at Cornell before being appointed to a lectureship in chemistry in York in 1965. He has been a full professor in York since 1983. His more than 250 publications are mainly in the area of vibrational spectroscopy, latterly focusing on time-resolved studies of photoreaction intermediates and on biomolecular systems in solution. He is active in environmental chemistry and is a founder member and former chairman of the Environment Group of The Royal Society of Chemistry and editor of 'Industry and the Environment in Perspective' (RSC, 1983) and 'Understanding Our Environment' (RSC, 1986). As a member of the Council of the UK Science and Engineering Research Council and several of its sub-committees, panels, and boards, he has been heavily involved in national science policy and administration. He was, from 1991–93, a member of the UK Department of the Environment Advisory Committee on Hazardous Substances and is currently a member of the Publications and Information Board of The Royal Society of Chemistry.

Roy M. Harrison, BSc, PhD, DSc (Birmingham), FRSC, CChem, FRMetS, FRSH

Roy M. Harrison is Queen Elizabeth II Birmingham Centenary Professor of Environmental Health in the University of Birmingham. He was previously Lecturer in Environmental Sciences at the University of Lancaster and Reader and Director of the Institute of Aerosol Science at the University of Essex. His more than 200 publications are mainly in the field of environmental chemistry, although his current work includes studies of human health impacts of atmospheric pollutants as well as research into the chemistry of pollution phenomena. He is a former member and past Chairman of the Environment Group of The Royal Society of Chemistry for whom he has edited 'Pollution: Causes, Effects and Control', (RSC, 1983; Second Edition, 1990) and 'Understanding our Environment: An Introduction to Environmental Chemistry and Pollution' (RSC, Second Edition, 1992). He has a close interest in scientific and policy aspects of air pollution, currently being Chairman of the Department of Environment Quality of Urban Air Review Group as well as a member of the DoE Expert Panel on Air Quality Standards and Photochemical Oxidants Review Group and the Department of Health Committee on the Medical Effects of Air Pollutants.

Contributors

Roger Atkinson, *Statewide Air Pollution Research Center, and Department of Soil and Environmental Sciences, University of California, Riverside, California 92521, USA*

Christophe Boissard, *Institute of Environmental & Biological Sciences, Lancaster University, Lancaster LA1 4YQ, UK*

Xu-Liang Cao, *Institute of Environmental & Biological Sciences, Lancaster University, Lancaster LA1 4YQ, UK*

J. Chandler, *Atmospheric Measurements and Processes Department, AEA Technology, National Environmental Technology Centre, E5 Culham, Abingdon, Oxfordshire OX14 3DB, UK*

Derrick R. Crump, *Building Research Establishment, Garston, Watford, Hertfordshire WD2 7JR, UK*

T. J. Davies, *Atmospheric Measurements and Processes Department, AEA Technology, National Environmental Technology Centre, E5 Culham, Abingdon, Oxfordshire OX14 3DB, UK*

M. Delaney, *Atmospheric Measurements and Processes Department, AEA Technology, National Environmental Technology Centre, E5 Culham, Abingdon, Oxfordshire OX14 3DB, UK*

Richard G. Derwent, *Atmospheric Processes Research Branch, Meteorological Office, London Road, Bracknell, Berkshire RG12 2SZ, UK*

G. J. Dollard, *Atmospheric Measurements and Processes Department, AEA Technology, National Environmental Technology Centre, E5 Culham, Abingdon, Oxfordshire OX14 3DB, UK*

S. Craig Duckham, *Institute of Environmental & Biological Sciences, Lancaster University, Lancaster LA1 4YQ, UK*

P. Dumitrean, *Atmospheric Measurements and Processes Department, AEA Technology, National Environmental Technology Centre, E5 Culham, Abingdon, Oxfordshire OX14 3DB, UK*

R. A. Field, *Atmospheric Measurements and Processes Department, AEA Technology, National Environmental Technology Centre, E5 Culham, Abingdon, Oxfordshire OX14 3DB, UK*

C. Nicholas Hewitt, *Institute of Environmental & Biological Sciences, Lancaster University, Lancaster LA1 4YQ, UK*

B. M. R. Jones, *Atmospheric Measurements and Processes Department, AEA Technology, National Environmental Technology Centre, E5 Culham, Abingdon, Oxfordshire OX14 3DB, UK*

Pauline M. Midgley, *M & D Consulting, Ludwigstraße 49, D-70771 Leinfelden, Germany*

John Murlis, *Air Quality Division, Department of the Environment, B 354, Romney House, 43 Marsham Street, London SW1P 3PY, UK*
(Present Address: Her Majesty's Inspectorate of Pollution, P3/015, 2 Marsham Street, London SW1P 3PY, UK)

P. D. Nason, *Atmospheric Measurements and Processes Department, AEA Technology, National Environmental Technology Centre, E5 Culham, Abingdon, Oxfordshire OX14 3DB, UK*

Neil R. Passant, *AEA Technology, National Environmental Technology Centre, Culham, Abindgon, Oxfordshire OX14 3DB, UK*

D. Watkins, *Atmospheric Measurements and Processes Department, AEA Technology, National Environmental Technology Centre, E5 Culham, Abingdon, Oxfordshire OX14 3DB, UK*

Sources, Distributions, and Fates of VOCs in the Atmosphere

RICHARD G. DERWENT

1 Introduction

Historical Background

The role and importance in atmospheric chemistry of organic compounds produced by human activity was established about fifty years ago by Haagen-Smit in his pioneering studies of Los Angeles smog.[1] He identified the key importance of hydrocarbon oxidation, in the presence of sunlight and oxides of nitrogen, as a photochemical source of ozone and other oxidants. Detailed understanding of the mechanism of photochemical smog formation has developed since then through the combination of smog chamber, laboratory chemical kinetics, field experiment, air quality monitoring, and computer modelling studies.

An understanding of the importance of the organic compounds emitted from the natural biosphere developed somewhat later with the recognition of the importance of the isoprene and terpene emissions from plants and trees.[2] The oxidation of these organic compounds leads to the production of carbon monoxide[3] and aerosol particles, the latter being responsible for the haze associated with forested regions.

Since these early pioneering studies, photochemical smog has subsequently been detected in almost all of the world's major urban and industrial centres, at levels which exceed internationally agreed criteria values set to protect human health.[4] Chlorinated organic compounds from human activities now reach the stratosphere, where processing by solar radiation yields active odd-chlorine species which are potent depleting agents of the stratospheric ozone layer.[5]

Despite the importance given now to organic compounds, their routine measurement in the atmosphere has only recently become commonplace. Furthermore, there are few detailed emission inventories for the major urban and industrial centres for which man-made emissions are fully resolved by species.

[1] A. J. Haagen-Smit, C. E. Bradley, and M. M. Fox, *Ind. Eng. Chem.*, 1953, **45**, 2086.
[2] R. A. Rasmussen and F. W. Went, *Proc. Natl. Acad. Sci. USA*, 1965, **53**, 215.
[3] E. Robinson and R. C. Robbins, SRI Project PR 6755, Stanford Research Institute, California, 1968.
[4] World Health Organisation, 'Air quality guidelines for Europe', European Series no. 23, WHO Regional Publications, Copenhagen, 1987.
[5] World Meteorological Office, 'Scientific Assessment of Ozone Depletion: 1991', Global Ozone Research and Monitoring Project Report no. 25, Geneva, Switzerland, 1992.

There is much research to be completed into the sources, distributions, and fates of organic compounds before photochemical smog control programmes can deliver the required air quality standards and before the role of organic compounds in the greenhouse effect is fully quantified.

Definitions

Volatile organic compounds, or VOCs, are an important class of air pollutants, commonly found in the atmosphere at ground level in all urban and industrial centres. There are many hundreds of compounds which come within the category of VOCs and the situation is yet further complicated by different definitions and nomenclature. Strictly speaking, the term volatile organic compounds refers to those organic compounds which are present in the atmosphere as gases, but which under normal conditions of temperature and pressure would be liquids or solids. A volatile organic compound is by definition an organic compound whose vapour pressure at say 20 °C is less than 760 torr (101.3 kPa) and greater than 1 torr (0.13 kPa). Many common and important organic compounds would be ruled out of consideration in this review if the upper and lower limits were adhered to rigidly.

In this chapter, this strict definition is not applied and the term VOC is taken to mean any carbon-containing compound found in the atmosphere, excluding elemental carbon, carbon monoxide, and carbon dioxide. This definition is deliberately wide and encompasses both gaseous carbon-containing compounds and those similar compounds adsorbed onto the surface of atmospheric suspended particulate matter. These latter compounds are strictly semi-volatile organic compounds. The definition used here includes substituted organic compounds, so that oxygenated, chlorinated, and sulfur-containing organic compounds would come under the present definition of VOC.

Other terms used to represent VOCs are hydrocarbons (HCs), reactive organic gases (ROGs), and non-methane volatile organic compounds (NMVOCs). The use of common names for the organic compounds is preferred in this review since these are more readily understood by industry and more commonly used in the air pollution literature. IUPAC names are however provided in all cases where they differ significantly from the common names.

Sources

Organic compounds are present in the atmosphere as a result of human activities, arising mainly from motor vehicle exhausts, evaporation of petrol vapours from motor cars, solvent usage, industrial processes, oil refining, petrol storage and distribution, landfilled wastes, food manufacture, and agriculture.[6] Natural biogenic processes also give rise to substantial ambient concentrations of organic compounds and include the emissions from plants, trees, wild animals, natural forest fires, and anaerobic processes in bogs and marshes.[7]

[6] S.D. Piccot, J.J. Watson, and J.W. Jones, *J. Geophys. Res.*, 1992, **97**, 9897.

[7] T.E. Graedel, T.S. Bates, A.F. Bouwman, D. Cunnold, J. Dignon, I. Fung, D.J. Jacob, B.K. Lamb, J.A. Logan, G. Marland, P. Middleton, J.M. Pacyna, M. Placet, and C. Veldt, *J. Biogeochem. Cycles*, 1993, **7**, 1.

Concerns

Because of the very large number of individual air pollutants that come within the above definition, their importance as a class of ambient air pollutants has only recently become recognized. Progress has been slow because intensive air monitoring to confirm their occurrence in the ambient atmosphere has only recently been started and because of the lack of basic information with which to target research activities. The situation has improved dramatically over the last few years and the important role played by organic compounds in a range of environmental problems of concern can now be identified.

These important roles are in:

- stratospheric ozone depletion
- ground level photochemical ozone formation
- toxic or carcinogenic human health effects
- enhancing the global greenhouse effect
- accumulation and persistence in the environment

These phenomena are briefly reviewed in the paragraphs below and some are discussed in more detail in the sections which follow.

Stratospheric Ozone Depletion. Many organic compounds are stable enough to persist in the atmosphere, to survive tropospheric removal processes, and to reach the stratosphere. If they contain chlorine or bromine substituents, the processes of stratospheric photolysis and hydroxyl radical destruction may lead to the release of active ozone-destroying chain carriers and to further stimulation of stratospheric ozone layer depletion and Antarctic 'ozone hole' formation.[5] Many chlorinated solvents and refrigerants, and bromine–containing fire retardants and fire extinguishers have been identified as belonging to the category of organic compounds which may lead to stratospheric ozone layer depletion. Such compounds come within the scope and control of the Montreal Protocol.[8]

Ground Level Ozone Formation. Organic compounds play a crucial role in ground level photochemical oxidant formation since they control the rate of oxidant production in those areas where NO_x levels are sufficient to maintain ozone production. The term 'hydrocarbons' is widely used in this context to refer to those organic compounds which take part in photochemical ozone production.[9] The contribution that organic compounds make to the exceedence of environmental criteria for ozone across Europe is now widely recognized. Long-range transboundary transport of ozone and action to control its precursors is an important feature of the problem. Organic compounds, which as a class produce photochemical ozone in the troposphere, come within the scope of the Geneva

[8] United Nations Environment Programme, 'Montreal Protocol on the Protection of the Ozone Layer', Nairobi, Kenya, 1992.
[9] J. M. Huess and W. A. Glasson, Hydrocarbon reactivity and eye irritation, *Environ. Sci. Technol.*, 1968, **2**, 1109.

Protocol to the UN ECE International Convention on Long Range Transboundary Air Pollution.[10]

Ground level ozone is of concern not only with respect to human health but also because of its effects on crops, plants, and trees. Elevated ozone concentrations during summertime photochemical pollution episodes may exceed environmental criteria[4] set to protect both human health and natural ecosystems.[11] It is these concerns which led to the formulation of the Geneva Protocol[10] and which underpin the reductions in emissions and control actions which it stipulates.

Toxic and Carcinogenic Health Effects. Organic compounds may have important impacts on human health through direct mechanisms in addition to their indirect impacts through photochemical ozone formation. Some organic compounds affect the human senses through their odour, some others exert a narcotic effect, and certain species are toxic.[4] Concern is particularly expressed about those organic compounds which could induce cancer in the human population: the human genotoxic carcinogens.[12] The term 'air toxics' is usually given to those organic compounds that are present in the ambient atmosphere and have or are suspected to have the potential to induce cancer in the human population.

The control of air toxics is currently both a national and an international activity, involving a wide range of international forums. A wide range of chemicals are also coming under scrutiny in this context. The most important organic compounds which belong to the air toxic category, and are widely distributed in the ambient atmosphere, include:

- benzene and 1,3-butadiene (buta-1,3-diene), as potential leukaemia-inducing agents
- formaldehyde (methanal), as a potential nasal carcinogen
- polynuclear aromatic hydrocarbons, as potential lung cancer inducing agents
- polychlorinated biphenyl compounds (PCBs) and polychlorinated terphenyl compounds (PCTs)
- dioxins and furans

Global Greenhouse Effect. Almost all of the organic compounds emitted as a result of human activities are emitted into the atmospheric boundary layer, the shallow region of the troposphere next to the earth's surface whose depth is typically a few hundred metres in winter to perhaps 2 km in mid-summer. Many of the reactive organic compounds are quickly oxidized in the atmospheric boundary layer. However, some survive and are transported into the free troposphere above the boundary layer during particular meteorological events

---

[10] United Nations Economic Commission for Europe, 'Protocol to the 1979 Convention on long-range transboundary air pollution concerning the control of emissions of volatile organic compounds or their transboundary fluxes', ECE/EB.AIR/30, Geneva, Switzerland, 1991.

[11] J. Fuhrer and B. Acherman, 'Critical Levels for Ozone', Schriftenreihe der FAC Liebefeld Nummer 16, Swiss Federal Research Station for Agricultural Chemistry and Environmental Hygiene, Liebefeld–Bern, Switzerland, 1994.

[12] United States Environmental Protection Agency, 'Cancer Risk from Outdoor Exposure to Air Toxics', USEPA OAQPS, Research Triangle Park, North Carolina, USA, 1990.

such as the passage of fronts, convection, and in the passage of air masses over mountains.

Some of the longer-lived organic compounds are accumulating in the troposphere, or may have the potential to do so. If any of these compounds can absorb solar or terrestrial infrared radiation, then they may contribute to the enhanced greenhouse effect. Such compounds would be classed as radiatively active gases and their relative effectiveness compared with carbon dioxide can be expressed through their Global Warming Potentials (GWPs – see page 105).[13]

Many organic compounds are not themselves radiatively active gases, but they do have the property of potentially being able to perturb the global distributions of other radiatively active gases. If they exhibit this property, then they can be classes as secondary greenhouse gases and indirect GWPs may be defined for them.[14] Organic compounds can behave as secondary greenhouse gases by:

- reacting to produce ozone in the troposphere
- increasing or decreasing the tropospheric ˙OH distribution and hence perturbing the distribution of methane

Once in the free troposphere, long-lived organic compounds can stimulate ozone production there. Ozone levels in this region are believed to be rising steadily[15] and this is of some concern because ozone is an important global greenhouse gas. However, the importance of the emissions of organic compounds from human activities in the global tropospheric ozone increase is still under evaluation.[13]

Accumulation and Persistence. Some of the higher molecular mass organic compounds are persistent enough to survive oxidation and removal processes in the boundary layer and may be transported over large distances before being removed in rain.[16] There is an important class of organic compounds, the semi-volatile VOCs which, because of their molecular size and complexity, tend to become adsorbed onto the surface of suspended particulate matter. In this form they undergo long-range transport and may be removed in rain remote from their point of original emission. Once deposited in rain, they may re-evaporate back into the atmosphere and begin the cycle all over again. Ultimately this material may be recycled through the atmosphere before reaching its more permanent sink in the colder aquatic environments in polar regions. Biological accumulation in these sensitive environments can lead to toxic levels in human foodstuffs in areas exceedingly remote from the point of original emission.

The identification of those organic compounds which are likely to persist in the environment, to bio-accumulate, and hence to find a pathway back to man, is still in its early days. Already some classes of organic compounds can be identified including the PCBs, PCTs, and phthalic acid and its derivatives. International

[13] J.T. Houghton, G.J. Jenkins, and J.J. Ephraums, 'Climate Change: The IPCC Scientific Assessment', Cambridge University Press, Cambridge, UK, 1990.

[14] R.G. Derwent, in 'Non-CO_2 Greenhouse Gases. Why and How to Control?', Kluwer Academic Publishers, Dordrecht, The Netherlands, 1994, p. 289.

[15] A. Volz and D. Kley, *Nature (London)*, 1988, **332**, 240.

[16] M. Oehme, *The Science of the Total Environment*, 1991, **106**, 43.

Table 1 Emissions of volatile organic compounds from both human activities and natural biogenic sources for each European country in thousand tonnes yr^{-1} (1989 figures)

Country	Man-made	Natural
Albania	33	10
Austria	441	37
Belgium	340	39
Bulgaria	167	203
Czechoslovakia	275	88
Denmark	125	5
Finland	181	72
France	1972	641
German Democratic Republic (now eastern Germany)	110	31
German Federal Republic (now western Germany)	2042	118
Greece	358	31
Hungary	358	113
Iceland	8	0
Ireland	124	3
Italy	2793	72
Luxembourg	10	2
Netherlands	473	9
Norway	245	27
Poland	982	77
Portugal	162	46
Romania	386	219
Spain	936	175
Sweden	460	104
Switzerland	304	7
Turkey	263	283
USSR (European part)	9064	2256
United Kingdom	1777	24
Yugoslavia	300	66

Sources:
[1] D. Simpson, *Atmos. Environ.*, 1993, **27A**, 921.
[2] A. Guenther, C. N. Hewitt, D. Erickson, R. Fall, C. Geron, T. Graedel, P. Harley, L. Klinger, M. Lerdau, W. A. McKay, T. Pierce, B. Scholes, R. Steinbrecher, R. Tallamraju, J. Taylor, and P. Zimmerman, *J. Geophys. Res.*, 1995, **100**, 8873.

action has yet to begin to tackle the problems of the long-range transboundary transport of compounds which may persist and accumulate in polar environments.

2 Sources of VOCs

Emission Inventories for European Countries

Emission inventories are now becoming available for the low molecular weight organic compounds for most European countries[17] and emission estimates are shown in Table 1 for 1989. The countries with the largest emissions appear to be

[17] D. Simpson, *Atmos. Environ.*, 1993, **27A**, 921.

USSR, Italy, and the Federal Republic of Germany. The major source categories identified include mobile sources through all modes of transport, stationary sources including evaporation, solvent usage, the industrial processes of oil refining and chemicals manufacture, oil and gas production, and agriculture.

Altogether, European emissions of low molecular mass volatile organic compounds from human activities amounted to about 23.8 million tonnes yr^{-1} in 1989.[17] This total is comparable with that of sulfur dioxide (as S) and nitrogen oxides (as NO_2), with each of the order of 20 million tonnes yr^{-1} for Europe as a whole.

Estimated emissions from natural sources are also included in Table 1. The latter are largely thought to be isoprene emissions from deciduous trees.[18] Natural emissions of isoprene appear to be somewhat lower, 4.8 million tonnes yr^{-1}, in total compared with that from man-made sources over Europe as a whole. Emissions from human activities appear to overwhelm natural sources in most countries. However, in some countries the reverse is true, *e.g.* in Bulgaria and Turkey, natural sources of isoprene predominate. In the United Kingdom, emissions of volatile organic compounds from human activities are about 80 times higher than those of isoprene from natural sources. Atmospheric VOCs from natural sources are discussed in more detail in Nicholas Hewitt's article on page 17.

Methane Emissions in the UK

In the United Kingdom, emissions of methane are subject to large uncertainties and have shown a slight downwards trend over the period 1970 to 1992.[19] An overall decrease in emissions from coal mines over the period has been largely offset by increases from gas leakage, landfill, and offshore oil and gas operations. In 1992, the total UK emissions of methane have been estimated as 4.7 million tonnes yr^{-1}. Animals account for about 30% of total emissions, with the largest contribution being from cattle.

Emissions of Other Organic Compounds in the UK

Emissions of organic compounds (methane excluded) in the UK for 1992 have been reported as 2.6 million tonnes yr^{-1}. A slight (10%) upwards trend in these emissions over the period since the 1970s has been documented.[19] Although there have been improvements in the accuracy of such emission estimates, there still remain substantial (30%) uncertainties.[20] In 1990, road transport accounted for 41% of the total, with chemical processes and solvents accounting for 50%.

By combining figures for the total emissions by source category with the profile

[18] A. Guenther, C. N. Hewitt, D. Erickson, R. Fall, C. Geron, T. Graedel, P. Harley, L. Klinger, M. Lerdau, W. A. McKay, T. Pierce, B. Scholes, R. Steinbrecher, R. Tallamraju, J. Taylor, and P. Zimmerman, *J. Geophys. Res.*, 1995, **100**, 8873.

[19] 'Digest of Environmental Protection and Water Statistics', Her Majesty's Stationery Office (HMSO), London, vol. 16, 1994.

[20] H. S. Eggleston, 'Accuracy of national air pollution emission inventories', Warren Spring Laboratory Report LR 715(AP), Stevenage, UK, 1991.

of the mass emissions of individual organic compounds, it is possible to derive national emission estimates for over 90 individual organic compounds.[21] These are shown in Table 2 for the UK summed over all source categories and presented as percentages of the total emissions.

On this basis, the speciated emissions of over 90 individual organic compounds have been identified in UK source categories. n-Butane (butane) appears to account for the greatest percentage, about 7% of the total. Of all the classes of VOC species, the alkanes appear to account for the greatest percentage of UK national emissions.

3 Ambient Concentrations of Organic Compounds

A summary is provided in Table 3 of the measured concentrations of organic compounds at six representative sites along a pollution gradient across Europe. The concentrations steadily decrease through three decades, from the urban kerbside, to urban background, to rural and to remote maritime background sites. Since the concentration ratios do not stay constant over the six sites, it is clear that the air at the remote sites is not merely diluted urban air. Many different sources, as well as depletion by chemical conversion, contribute to the observed spatial patterns and the differences between the mean concentrations at the different sites.

The species distribution of the organic compounds at the remote maritime sites are dominated by two paraffins (alkanes), ethane and propane, presumably reflecting the importance of natural, marine sources. By comparison, the species distribution observed at the urban kerbside site is heavily dominated by ethylene (ethene), n-butane (butane), and acetylene (ethyne), the major components of motor vehicle exhaust.

4 Fates of Organic Compounds

As a class, volatile organic compounds all share the same major atmospheric removal mechanisms, which include the following (see also the article by Roger Atkinson on page 65):

- photochemical oxidation by hydroxyl (˙OH) radicals in the troposphere
- photolysis in the troposphere and stratosphere
- deposition and uptake at the earth's surface
- reaction with other reactive species such as chlorine atoms, nitrate radicals at night, and ozone

Photochemical Oxidation by ˙OH Radicals in the Troposphere

The reactive free radical species, hydroxyl or ˙OH, plays a central role in tropospheric chemistry[22] by cleansing the atmosphere of most of the trace gases emitted by terrestrial processes and by human activities, particularly the organic

[21] United Kingdom Photochemical Oxidants Review Group (PORG), 'Ozone in the United Kingdom: 1993', Harwell Laboratory, UK, 1993.
[22] H. Levy, *Science*, 1971, **173**, 141.

compounds which are the subject of this review. This steady state of hydroxyl radicals is maintained by a set of rapid free radical reactions which comprise the fast photochemical balance of the troposphere and so define the oxidation capacity of the troposphere.[23]

The hydroxyl radical oxidation sink for organic compounds is operating throughout the troposphere and is not limited in its spatial regime merely to the atmospheric boundary layer. This oxidation sink determines the atmospheric lifetimes for the vast majority of this class of atmospheric species. Lifetimes, together with emission rates, determine the global concentrations which would eventually build up if emissions continued indefinitely.

It is difficult to generalize about the atmospheric lifetimes for a wide range of organic compounds which result from oxidation by tropospheric hydroxyl radicals.[24] The paraffins (alkanes) generally have lifetimes of about 2–30 days, with lifetime decreasing along the homologous series. The first two members of the alkane series have significantly longer lifetimes than the above range, with methane about 10 years and ethane about 120 days. Olefins (alkenes) generally have the shortest lifetimes of the major classes of organic compounds found in the atmosphere. Their lifetimes range from about 0.4–4 days, with lifetimes decreasing along the homologous series. Aromatic hydrocarbons show lifetimes in the range 0.4–5 days, with the first member of that series, benzene, showing an uncharacteristically long lifetime of 25 days.

Photolysis in the Troposphere and the Stratosphere

Photolysis is an important removal process for only the limited range of organic compounds which show strong absorption features in the ultraviolet and visible regions of the spectrum. The extent of the overlap between these absorption features and the solar spectrum, the quantum yields for the various pathways and the solar actinic flux determine the lifetime of the photochemically labile species.[25] The solar actinic fluxes vary considerably with time-of-day, latitude, season, and quite importantly with height in the atmosphere.[26] The solar spectrum seen by photochemically labile organic compounds is characteristically different in the troposphere and stratosphere. In the latter, the wavelengths extend down to 180 nm, whereas in the former they do not extend much below 295 nm.

Photolysis is an important loss process for the aldehydes and ketones in the troposphere where it also acts as an important source of free radical species.[22] Photolysis lifetimes of aldehydes and ketones in the sunlit troposphere may be of the order of several days. In the stratosphere, vacuum ultraviolet photolysis of chlorine-containing organic compounds is an important removal mechanism for the chlorocarbons. This latter process, however, acts as a source of active chlorine carriers which can catalyse the destruction of the ozone layer in the presence of polar stratospheric clouds.[5] Lifetimes due to stratospheric photolysis

[23] P. J. Crutzen, *Tellus*, 1974, **26**, 47.

[24] R. G. Derwent, *Phil. Trans. R. Soc. Lond.*, 1995, **A351**, 205.

[25] A. M. Hough, 'The calculation of photolysis rates for use in global tropospheric modelling studies, AERE Report AERE–R13259, Her Majesty's Stationery Office (HMSO), London, 1988.

[26] K. L. Demerjian, K. L. Schere, and J. T. Peterson, *Adv. Environ. Sci. Technol.*, 1980, **10**, 369.

Table 2 UK emissions of volatile organic compounds. Percentage (by mass distribution) of different species emitted in 1990

Common name	IUPAC name	Percentage by mass (%)
ethane		1.68
propane		0.53
n-butane	butane	6.5
n-pentane	pentane	2.1
isopentane	2-methylbutane	3.4
n-hexane	hexane	1.4
2-methylpentane		1.2
3-methylpentane		0.80
2,2-dimethylbutane		0.27
2,3-dimethylbutane		0.36
n-heptane	heptane	0.35
2-methylhexane		0.54
3-methylhexane		0.47
n-octane	octane	0.29
2-methylheptane		1.6
n-nonane	nonane	1.1
2-methylnonane		1.1
n-decane	decane	1.1
n-undecane	undecane	1.2
n-dodecane	dodecane	0.33
cyclohexane		0.0013
methylcyclohexane		0.21
ethylene	ethene	3.8
propylene	propene	1.6
1-butene	but-1-ene	0.49
2-butene	but-2-ene	0.94
butylene	2-methylpropene	0.22
1-pentene	pent-1-ene	0.32
2-pentene	pent-2-ene	0.62
2-methylbut-1-ene		0.12
3-methylbut-1-ene		0.16
2-methylbut-2-ene		0.27
styrene	ethenylbenzene	0.36
isoprene	2-methylbuta-1,3-diene	0.13
acetylene	ethyne	1.7
benzene		2.3
toluene	methylbenzene	2.3
o-xylene	1,2-dimethylbenzene	2.6
m-xylene	1,3-dimethylbenzene	3.1
p-xylene	1,4-dimethylbenzene	3.0
ethylbenzene		1.3
n-propylbenzene	propylbenzene	0.5
cumene	2-methylethylbenzene	0.41
1,2,3-trimethylbenzene		0.54
1,2,4-trimethylbenzene		1.2
1,3,5-trimethylbenzene		0.61
o-ethyltoluene	1-ethyl-2-methylbenzene	0.57
m-ethyltoluene	1-ethyl-3-methylbenzene	0.72
p-ethyltoluene	1-ethyl-4-methylbenzene	0.72
1,3-dimethyl-5-methylbenzene		0.49

Table 2 *continued*

Common name	IUPAC name	Percentage by mass (%)
1,3-diethyl-5-methylbenzene		0.49
formaldehyde	methanal	0.78
acetaldehyde	ethanal	0.15
propanal		0.16
butanal		0.10
methylpropanal		0.09
pentanal		0.016
benzaldehyde	benzene carbanal	0.073
acetone	propanone	0.89
methyl ethyl ketone	butanone	0.80
methyl isobutyl ketone	2-methylpentan-2-one	1.4
cyclohexanone		0.54
methyl alcohol	methanol	0.017
ethyl alcohol	ethanol	4.3
isopropanol	2-methylethanol	0.41
n-butanol	butanol	1.4
isobutanol	2-methylpropanol	1.0
s-butanol	butan-2-ol	0.87
t-butanol	2,2-dimethylethanol	0.0013
cyclohexanol		0.21
diacetone alcohol	4-methyl-4-hydroxypentan-2-one	0.87
dimethyl ether		0.0013
methyl t-butyl ether		0.0013
methoxypropanol		0.23
butyl glycol		0.86
methyl acetate	methyl ethanoate	0.0018
ethyl acetate	ethyl ethanoate	0.73
n-propyl acetate	propyl ethanoate	0.0013
isopropyl acetate	2-methylethyl ethanoate	0.0027
n-butyl acetate	butyl ethanoate	0.44
isobutyl acetate	2-methylpropyl ethanoate	0.44
formic acid	methanoic acid	0.0021
acetic acid	ethanoic acid	0.0021
propionic acid	propanoic acid	0.0013
methyl chloride	chloromethane	0.039
methylene chloride	dichloromethane	0.40
methyl chloroform	1,1,1-trichloroethane	1.1
tetrachloroethylene	tetrachloroethene	0.79
trichloroethylene	trichloroethene	1.1
cis-dichloroethylene	*cis*-dichloroethene	0.0013
trans-dichloroethylene	*trans*-dichloroethene	0.0013
vinyl chloride	chloroethene	0.24

Sources:
[1] United Kingdom Photochemical Oxidants Review Group, 'Ozone in the United Kingdom: 1993', Harwell Laboratory, UK, 1993.
[2] R. G. Derwent, M. E. Jenkin, and S. M. Saunders, *Atmos. Environ.*, submitted.

for organic compounds released at the earth's surface are generally of the order of 40 years or more.

Deposition and Uptake at the Earth's Surface

Deposition onto water surfaces, plants, vegetation, and soil surfaces is generally termed dry deposition, and requires both the transport of the trace gas species to the surface within the atmospheric boundary layer and its subsequent reaction or adsorption at the surface or on surface elements.[27] Dry deposition, therefore, only tends to act efficiently on those organic compounds present in the atmosphere close to the surface where biological uptake occurs.

For the majority of the organic compounds in this review, little information is available concerning the importance of dry deposition. The general impression is that this process is not important. For example, soil uptake of methane accounts for a removal lifetime of 160 years.[13]

The removal of trace gases by precipitation, referred to as wet deposition, results from the incorporation of material into falling precipitation (wash-out) and by incorporation into cloud droplets (rain-out). These removal processes are necessarily only significant for those species which are readily soluble.[28] The vast majority of low molecular mass organic compounds are not in this category and so are generally not removed significantly by wet deposition. The highly polar carboxylic acids and alkyl hydroperoxides are probably the only classes of organic compounds which undergo wet removal.

As the molecular mass of an organic compound increases, its volatility tends to decrease and increasingly it becomes adsorbed onto the atmospheric aerosol. Semi-volatile organic compounds tend to behave more like aerosol particles than the parent organic compounds from which they are formed. Deposition by dry and wet deposition rather than oxidation by hydroxyl radicals are the major removal mechanisms for semi-volatile organic compounds.

Reactions with Chlorine Atoms, Nitrate Radicals, and Ozone

Of all the reactive atoms and radical species, the hydroxyl radical is relatively unusual in its high reactivity with most inorganic and organic substances found in the atmosphere.[29] Fluorine atoms share this wide spectrum of reactivity with the hydroxyl radical but lack significant atmospheric sources. Chlorine atoms react rapidly with most organic compounds but, like fluorine, lack significant atmospheric sources. Except in rather special circumstances, chlorine atom reactions can generally be neglected in the determination of atmospheric lifetimes of organic compounds.

During night-time in the unpolluted troposphere, a steady state of nitrate radicals builds up through the reactions of nitrogen dioxide with ozone. Nitrate radicals may react with the highly reactive alkenes and dialkenes to form nitrato-carbonyl compounds by addition reactions.[29] The atmospheric fates of

[27] B. B. Hicks, *Water, Air Soil Pollut.*, 1986, **30**, 75.
[28] J. M. Hales, in 'The Handbook of Environmental Chemistry', Springer-Verlag, Berlin, 1986, p. 149.
[29] R. Atkinson, *J. Phys. Chem. Ref. Data*, Monograph 2, 1994, 1.

Table 3 Annual mean concentrations of organic compounds in ppt, measured at various locations in Europe

Organic compound		Izana (Canaries)[a] remote maritime	Rorvik (Sweden)[b] remote maritime	Langenbrugge (Germany)[c] remote rural	West Beckham (UK)[d] remote rural	Middlesbrough (UK)[e] urban background	Exhibition Road London (UK)[f] urban roadside
Common name	IUPAC name						
ethane	ethane	675	1707	2191	2804	2300	7375
ethylene	ethene	35	598	791	1546	1600	10825
propane	propane	115	732	1039	1313	4350	3525
propylene	propene	7	95	188	270	2100	5800
isobutane	2-methylpropane	33	313	286	722	2200	6375
acetylene	ethyne	65	767	1053	1396	6550	19300
n-butane	butane	22	545	540	1637	4350	12225
sum butenes				265		2450	3600
cyclopentane				22			
isopentane	2-methylbutane		328	265	855	2100	7450
propyne				22		260	300
n-pentane	pentane		241	170	443	500	2575
1,3-butadiene	buta-1,3-diene			20		650	800
sum pentenes				45		350	870
unresolved C6				115			
sum isohexanes	sum methylpentanes			149		200	850
isoprene	2-methylbuta-1,3-diene			19			200
n-hexane	hexane			105	144	200	600
sum methylhexanes	sum methylhexanes			130			
n-heptane	heptane			77		200	
sum isoheptanes							
benzene		30		351	725	1150	4625
toluene	methylbenzene	40		441	999	1550	7475
ethylbenzene				49	244	1100	1175
m- + p-xylene	sum dimethylbenzenes			140	537	650	3825
o-xylene	1,2-dimethylbenzene			62	242	450	1450
1,2,4-trimethylbenzene							300
1,3,5-trimethylbenzene							1075

[a] R. Schitt and P. Matusca, in 'Photo-oxidants: precursors and products', SPB Academic Publishing bv, Den Haag, The Netherlands, 1992, p. 131.
[b] A. Lindskog and J. Moldanova, *Atmos. Environ.*, 1991, **28**, 2383.
[c] S. Solberg, N. Schmidbauer, U. Pedersen, and J. Schaug, 'VOC measurements August 1992 – June 1993', EMEP/CCC-Report 6/93, Norwegian Institute for Air Research, Lillestrom, Norway, 1993.
[d] Photochemical Oxidants Review Group, 'Ozone in the United Kingdom', Department of the Environment, London, 1993.
[e] J. Derwent, P. Dumitrean, J. Chandler, T. J. Davies, R. G. Derwent, G. J. Dollard, M. Delaney, B. M. R. Jones, and P. D. Nason, 'A preliminary analysis of hydrocarbon monitoring data from an urban site'. AEA CS 18358030/005/issue2, AEA Technology, Harwell Laboratory, Oxfordshire, 1994.
[f] R. G. Derwent, D. R. Middleton, R. A. Field, M. E. Goldstone, J. N. Lester, and R. A. Perry, *Atmos. Environ.*, 1995, **29**, 923.

the various bifunctional addition compounds have yet to be identified. For the large number of organic compounds, however, nitrate radicals are not reactive enough to contribute significantly to atmospheric removal.

Ozone reactions appear to be significant compared with hydroxyl radical reactions for this same class of alkenes and dialkenes as for nitrate radicals.[29] Atmospheric lifetimes for the monoterpene natural biogenic hydrocarbons, whose reactions with ozone are significant, are found to be in the range of hours rather than days.

Fates

The overall impact of these removal processes on the fates and behaviour of the organic compounds emitted into the atmosphere by human activities and by natural sources is markedly dependent upon the physical and chemical properties of the individual organic compound. For the bulk of the organic compounds emitted by human activities in the northern mid-latitudes, atmospheric lifetimes are generally one hundred days or less.[30] They are likely to spread vertically up to the tropopause and through much of the northern hemisphere, at least over the continental regions. For those with lifetimes of five days or less, they are likely to be found to any significant extent only in the atmospheric boundary layer and within a thousand kilometres or so of major source regions.

Highly unreactive organic compounds may exhibit atmospheric lifetimes measured in years or tens of years. An important class is that of the ozone-depleting substances, where lifetimes span 5 years for methyl chloroform (1,1,1-trichloroethane) to about 130 years for CFC-12.[5] Some organic compounds with long atmospheric lifetimes are also important radiatively-active gases. This class includes methane with its 10 year lifetime and the replacement CFCs such as HFC 134a and 143a with lifetimes of 16 and 41 years, respectively.[13]

For those low-volatility high-molecular mass organic compounds, their fate is largely determined by the extent of their attachment by adsorption to the atmospheric aerosol.[16] Semi-volatile organic compounds attached to aerosol particles behave quite differently from gaseous organic compounds. The former are removed from the atmosphere largely by wet and dry deposition and the latter by hydroxyl radical oxidation. Lifetimes for semi-volatile organic compounds adsorbed onto aerosol particles are similar to those of aerosol particles themselves, generally about 5–10 days, close to the Earth's surface. Lifetimes for gaseous organic compounds are highly variable from days to years.

Fates are different also for the gaseous organic compounds compared with the organic compounds adsorbed onto particles. Atmospheric oxidation involves the complete destruction of the organic compound, ultimately to carbon dioxide and water, whereas deposition of the semi-volatile organic compounds leads to ecosystem contamination and transfer of the organic compound into different environmental media.

---

[30] R.G. Derwent, *Phil. Trans. Proc. Roy. Soc. A*, submitted.

5 Acknowledgements

The author is grateful to Robert Field and to Geoff Dollard of the National Environmental Technology Centre, Culham Laboratory, for making hydrocarbon data available for Exhibition Road, London, and for Middlesbrough, respectively. Support from the Department of the Environment Air Quality Research Programme through contract no. EPG 1/3/16 is acknowledged.

Atmospheric VOCs from Natural Sources

C. NICHOLAS HEWITT, XU-LIANG CAO,
CHRISTOPHE BOISSARD, AND S. CRAIG DUCKHAM

1 Introduction

Episodes of high concentrations of ozone in the lower troposphere occur over all parts of Europe and North America every summer, with even the northern and maritime countries being adversely affected. There is also evidence that the concentrations and frequency of ozone episodes are increasing,[1-2] and that ozone concentrations at rural sites in the UK are increasing over time.[3] Ozone is phytotoxic, with effects on cell permeability, leaf necrosis, crop yield reductions, and a possible involvement in forest decline all documented.[4] It is formed in the troposphere by photochemical reactions involving reactive hydrocarbons in the presence of oxides of nitrogen and sunlight, and these hydrocarbons may be of anthropogenic origin or may be naturally produced by the biosphere. Isoprene (C_5H_8) and a range of monoterpene $(C_{10}H_{16})$ compounds are considered the most important species in this latter category, although it is known that many other organic compounds, including acids, alcohols, esters, and ketones, are emitted by the biosphere. There is therefore considerable interest in quantifying the emission rates of isoprene and the monoterpenes to the atmosphere. Indeed, without this information it is likely that cost-effective strategies cannot be formulated for controlling the formation of ozone and other photochemically produced secondary air pollutants in the troposphere.

2 Measurement Methods

Sampling and Analytical Methods

Several different methods are used for the sampling of gaseous components in air. Each method has its own range of application, and it is important that suitable sampling techniques should be used. The two principal methods used for the determination of VOCs from biogenic sources are the whole-air and the adsorption techniques.

[1] A. Volz and D. Kley, *Nature (London)*, 1988, **332**, 240.

[2] A.M. Hough and R.G. Derwent, *Nature (London)*, 1990, **344**, 645.

[3] 'Ozone in the United Kingdom 1993', Third report of the United Kingdom Photochemical Oxidants Review Group (PORG), Department of the Environment, London, 1993.

[4] N.M. Darrall, *Plant Cell Environ.*, 1989, **12**, 1.

(1) Whole-air sampling involves the direct collection and isolation of the test atmosphere in an impermeable container, and generally requires relatively simple equipment. This technique is limited to those gaseous constituents for which sensitive analytical techniques are available, or which have high concentrations in the atmosphere. It is ideal for the light hydrocarbons, but is in general not applicable to less volatile compounds due to their possible adsorptive losses on the walls of the sample containers.

(2) Sampling by pumping air through an adsorption tube packed with adsorbent(s), followed by thermal desorption, is the most widely used method for the sampling of VOCs at low concentrations in air. Suitable adsorbents should be used for the sampling of different hydrocarbons to ensure not only the representative collection of the hydrocarbons of interest, but also their subsequent complete desorption for analysis. The most commonly used adsorbent for this purpose is Tenax (GC or TA). It has the desired property of not retaining significant amounts of water, but its adsorption capacity for highly volatile hydrocarbons (fewer than six carbons) is poor, and it has problems of artifact formation by reaction with ozone in air.[5-6] It has been found that some hydrocarbons, including the reactive biogenic monoterpenes, can be partly or completely decomposed during thermal desorption on some adsorbents.[7-8]

(3) Because of the complexity of the mixture of hydrocarbon compounds present in air, an analytical method that can resolve one compound from another is required. Gas chromatography (GC), particularly combined with the use of a high resolution capillary column, offers excellent separation. Although other GC detectors, such as the photoionization detector (PID), may be used for the analysis of hydrocarbons, the most widely used are the mass spectrometer and the flame ionization detector (FID).

(4) The FID is traditionally considered a highly non-selective detector, and can respond to almost all volatile organic compounds. It has therefore been widely used for the determination of volatile organic compounds in air, and has undergone little change in the last two decades. One of the advantages of the FID is that its response factors for true hydrocarbons can be predicted from the number of carbon atoms in the molecule. Thus, concentrations of all other hydrocarbons can be determined from calibration with only a single hydrocarbon. This is not the case for other organic compounds containing oxygen, nitrogen and halogens, *etc.* The absolute practical detection limit for a typical flame system is about 50 pg.

(5) Mass spectrometry (MS) has been used for the identification of organic compounds, and many different monoterpenes and oxygenated compounds have been identified in vegetation emissions. MS is normally used for the identification of organic compounds, but it has also been occasionally used for both qualitative and quantitative analysis. The detection limits of the GC–MS systems depend on the mode of operation (scanning mode and selected ion monitoring mode), and

[5] J. M. Roberts, F. C. Fehsenfeld, D. L. Albritton, and R. E. Sievers, in 'Identification and Analysis of Organic Pollutants in Air', ed. L. H. Keith, Butterworth, London, 1984.

[6] X.-L. Cao and C. N. Hewitt, *Environ. Sci. Technol.*, 1993, **28**, 757.

[7] V. A. Isidorov, I. G. Zenkevich, and B. V. Ioffe, *Atmos. Environ.*, 1985, **19**, 1.

[8] X.-L. Cao and C. N. Hewitt, *Chemosphere*, 1993, **27**, 695.

are compound-specific, but are generally similar or slightly higher than that of the GC–FID.

Methods for sampling and analysis of oxygenated compounds (*e.g.* aldehydes and ketones) using DNPH cartridges are also well characterized.[9] Air samples, normally 10–100 litres, are collected on dinitrophenylhydrazine (DNPH) coated C-18 'Sep-pak' cartridges at around 0.5–2 litres min[−1] and hydrazone derivatives are eluted with acetonitrile. Separation of analytes is obtained using HPLC with a C-18 ODS column (5 μm × 250 mm × 4 mm) using a solvent gradient of acetonitrile, water, and tetrahydrofuran; detection is at 360 nm. Pre-coated sample cartridges can be obtained commercially or prepared in the laboratory. Alternative methods include the use of dansylhydrazine (DNSH) treated cartridges with fluorescence detection of derivatives, and sampling of formaldehyde with annular denuders.

Emission Flux Measurement Methods

The emission fluxes of biogenic hydrocarbons from vegetation can be measured by several different techniques, such as the bag enclosure method, the tracer technique, the gradient profile method, and the recently developed conditional sampling or relaxed eddy accumulation method. The enclosure technique is indirect and measures the hydrocarbon flux from a relatively small sample of plant material, whereas the other techniques determine the average compound flux from vegetation covering a large area (typically 10^5 m^2 or more).

Bag Enclosure. In a dynamic flow-through branch enclosure system,[10] a Teflon bag is placed around a branch of vegetation, and ambient air is pumped into the chamber. The emission rate (E, ng [g (dry wt)]$^{-1}$ h^{-1}) and the corresponding emissions flux (F, ng m^{-2} h^{-1}) can then be calculated according to the following equations:

$$E = \frac{(m_{out}/f_{out} - m_{in}/f_{in})F_{in}}{TM} \tag{1}$$

$$F = EB \tag{2}$$

where m_{out} and m_{in} are the masses (ng) of hydrocarbon in the outflow and inflow samples, f_{out} and f_{in} is the sampling flow rate (cm^3 min^{-1}) for the outflow and inflow, F_{in} is the total flow rate (cm^3 min^{-1}) into the chamber, T is the sampling time (h), M is the dry weight (g) of the leaves or needles on the branch, and B is the biomass factor (g m^{-2}) appropriate to the particular forest site.

This method is very simple and easy to perform, can sample different species individually, and does not require highly sensitive or fast response chemical detectors nor the meteorological data required by the other techniques. It can be employed in the field or in the laboratory where the effects of different

[9] S. B. Tejada, *Int. J. Environ. Chem.*, 1986, **26**, 167.
[10] R. A. Street, J. Wolfenden, S. C. Duckham, and C. N. Hewitt, in 'Photo-oxidants: precursors and products', ed. P. M. Borell *et al.*, SPB Academic Publishing, Den Haag, The Netherlands, 1993.

environmental conditions can be investigated systematically. Thus, this method has been widely used for the measurement of emission fluxes of hydrocarbons (mainly isoprene and monoterpenes) from vegetation[11-14] since its development by Zimmerman.[15,16] However, since physical confinement of the plant under investigation is required, the enclosure technique may perturb the normal biological function of the plant and hence yield unrepresentative emission flux rates. In addition, a detailed biomass survey is required to allow extrapolation from a single branch to a forest or region in order to use these measurements for calculation of an emission inventory.

Gradient Profile. This method is based upon micrometeorological surface layer theory, and can be used to obtain fluxes from plants distributed over much larger areas. The hydrocarbon concentration gradients (dC/dz) above an essentially infinite, uniform plane source (*e.g.* an ideal forest canopy) can be obtained by measuring hydrocarbon concentrations at several different heights. Temperature and wind speed or water vapour concentration gradients must also be measured. The meteorological data are used to determine the eddy diffusivity (K_z) so that the hydrocarbon emission flux can be calculated from the concentration gradient according to the following equation:

$$F = K_z \frac{dC}{dz} \tag{3}$$

However, since it is difficult to set up and has extremely stringent sensor and site requirements, this method has been used only occasionally, primarily as an independent check upon enclosure measurements.[11] It has also been found recently[13,14] that negative emission fluxes may result from the complex diurnal variations of meteorological conditions (*e.g.* the nocturnal temperature inversion). In addition, this method may also be affected by chemical reactions between hydrocarbons and the oxidizing species (*e.g.* O_3, and $^\bullet OH$ and $NO_3^\bullet$ radicals) during their upward transport, which may result in unrepresentative emission fluxes.

Tracer Technique. The tracer technique involves simulation of forest emissions by release of an inert tracer (*e.g.* SF_6), and measurement of ambient concentrations of the tracer and the biogenic hydrocarbons of interest along a downwind sample line. Lamb *et al.*[17] described the application of this technique to the determination of isoprene emission fluxes from an isolated grove of Oregon white oak. Based on the known tracer release rate and measured ambient concentrations of isoprene

[11] B. Lamb, H. Westberg, and G. Allwine, *J. Geophys. Res.*, 1985, **90**, 2380.

[12] R.W. Janson, *J. Geophys. Res.*, 1993, **98**, 2839.

[13] C. Boissard, C. N. Hewitt, R. A. Street, S. C. Duckham, X.-L. Cao, I. J. Beverland, D. H. O'Neill, B. J. Moncrieff, R. Milne, and D. Fowler, *J. Geophys. Res.*, 1995, in the press.

[14] C. N. Hewitt *et al.*, 1995, unpublished data.

[15] P. R. Zimmerman, Report EPA-450/4-70-004, US Environmental Protection Agency, Research Triangle Park, North Carolina, 1979.

[16] P. R. Zimmerman, Report EPA-904/9-77-028, US Environmental Protection Agency, Research Triangle Park, North Carolina, 1979.

[17] B. Lamb, H. Westberg, and G. Allwine, *Atmos. Environ.*, 1986, **20**, 1.

and tracer, isoprene emission fluxes were calculated by two methods. The simplest one involved only a consideration of the ratio of the observed maximum isoprene and tracer concentrations:

$$F_i = F_{tracer} \frac{C_i(\max)}{C_{tracer}(\max)} \tag{4}$$

where F_i is the calculated isoprene flux ($\mu g\, m^{-2}\, h^{-1}$), F_{tracer} is the known tracer flux ($\mu g\, m^{-2}\, h^{-1}$), $C_i(\max)$ is the maximum isoprene concentration and $C_{tracer}(\max)$ is the maximum tracer concentration observed along the sample line.

The tracer technique does not require perturbation of the vegetation nor does it rely upon precise gradient measurements. However, its application has been limited by two assumptions: (1) the tracer release network should provide an accurate simulation of the biogenic emission of hydrocarbons of interest, and (2) the biogenic hydrocarbons of interest should be conserved (*i.e.* there should be no chemical loss) between the emission source and downwind sample points. Thus, this method is suitable for nonreactive gases, such as methane, but may not be suitable for the highly reactive hydrocarbons (*e.g.* monoterpenes), especially under conditions of high concentrations of ozone and hydroxyl radicals.

Eddy Correlation, Eddy Accumulation, and Relaxed Eddy Accumulation Techniques. The most direct approach to flux measurement is the eddy correlation technique (EC). This technique is based on the mean product of the fluctuations of vertical wind velocity (w) and concentration of the gas of interest (c). However, since this technique requires continuous fast response sensors with sufficient resolution to accurately measure the covariances of vertical wind velocity and concentrations of the gas of interest, it has not been used for organic compounds to date.

The eddy accumulation technique (EA), proposed by Desjardins,[18] overcomes the need for fast response gas sensors without adding other uncertainties, since it is based on the same physical principles as EC. In this method, air is drawn from the immediate vicinity of an anemometer measuring vertical windspeed (w) and diverted into one of two 'accumulators' on the basis of the sign of w, at a pumping rate proportional to the magnitude of w. Gas samples can be collected from the two accumulators and analysed with a slow response detector.

Businger and Oncley[19] suggested that the demands of eddy accumulation might be 'relaxed' by sampling air at a constant rate for updrafts and downdrafts, rather than proportionally (relaxed eddy accumulation, REA, or conditional sampling). They proposed that the flux (F) of the compound of interest should be given by

$$F = \beta \cdot \sigma_w (c^+ - c^-) \tag{5}$$

where β is an empirical constant (about 0.6), σ_w is the standard deviation of the

[18] R. L. Desjardins, *Boundary Layer Meteorol.*, 1977, **11**, 147.
[19] J. A. Businger and S. P. Oncley, *J. Atmos. Oceanic Technol.*, 1990, **7**, 349.

vertical windspeed ($m\,s^{-1}$) and c^+ and c^- are the mean concentrations ($\mu g\,m^{-3}$) of the gas in the upward- and downward-moving eddies.

Due to their technical difficulties, the EA and REA techniques have mainly been used only to measure emission fluxes of nonreactive species such as methane, CO_2, and H_2O.[20–23] Efforts have recently been made to use these techniques to determine the emission fluxes of biogenic hydrocarbons from vegetation.[13–14]

3 Sources of Natural VOCs in the Atmosphere

Sources of Biogenic VOCs

Isoprene appears to be derived from five-carbon intermediates of the mevalonic pathway associated with isoprenoid biosynthesis. The isoprene emission rates are dependent on the activity of the isoprene synthase enzyme.[24] Monoterpenes are derives from isoprenoids synthesized in the chloroplasts of plants and their biosynthesis has been comprehensively reviewed by Croteau.[25] Synthesis is often, but not exclusively, within specialized secretory ducts which can act as a terpene pool within leaves. Indeed, a significant proportion of emitted monoterpenes may be derived from recent biosynthesis.

An exhaustive literature survey of the emission of VOCs from plants is available.[26] Generally, deciduous trees are mainly isoprene emitters and coniferous trees monoterpene emitters, though some plants are both isoprene and monoterpene emitters (*e.g.* Sitka spruce) or do not emit at all. In addition, many crop and grass species are emitters of isoprene and/or monoterpenes,[27] and may also emit a range of C_6 oxygenated biogenic VOCs (*e.g.* hexenyl acetate and *trans*-2-hexenol).[28]

In a study of Mediterranean vegetation, around 20 different tree and shrub species were screened for tendencies to emit VOCs using the bag enclosure method with analysis by GC–MS and GC–FID (authors' unpublished data). The plant species which produced most hydrocarbons (on a dry weight basis) were: *Quercus ilex* (Holm, or holly, oak), which emitted largely monoterpenes; *Pinus pinea* (pine) which produced mainly linalool and some terpenes; and *Erica arborea* (white, or tree, heath; or *bruyère*) and *Myrtus communis* (common myrtle) which were predominantly isoprene emitters. A total of 32 compounds were identified as biogenic emissions. *Q. ilex* and *P. pinea* showed the greatest diversity of emitted species, producing 24 of the 32 compounds identified, with a

[20] J.M. Baker, J.M. Norman, and W.L. Bland, *Agric. Forest Meteorol.*, 1992, **62**, 31.

[21] E. Pattey, R.L. Desjardins, and P. Rochette, *Boundary Layer Meteorol.*, 1993, **66**, 341.

[22] S.P. Oncley, A.C. Delany, T.W. Horst, and P.P. Tans, *Atmos. Environ.*, 1993, **27A**, 2417.

[23] I.J. Beverland, R. Milne, C. Boissard, D.H. O'Neill, J.B. Moncrieff, and C.N. Hewitt, *J. Geophys. Res.*, 1995, submitted.

[24] J. Kuzma and R. Fall, *Plant Physiol.*, 1993, **101**, 435.

[25] R. Croteau, *Chem. Rev.*, 1987, **87**, 929.

[26] C.N. Hewitt and R.A. Street, *Atmos. Environ.*, 1992, **26A**, 3069.

[27] S.A. Arey, A.M. Winer, R. Atkinson, S.M. Aschmann, W.D. Long, C.L. Morrison, and D.M. Olszyk, *J. Geophys. Res.*, 1991, **96**, 9329.

[28] S.A. Arey, A.M. Winer, R. Atkinson, S.M. Aschmann, W.D. Long, and C.L. Morrison, *Atmos. Environ.*, 1991, **25A**, 1063.

total assigned plant emission rate as high as $35\,\mu g\,[g\,(dry\,wt)]^{-1}\,h^{-1}$ under ambient conditions.

In a study of the most dominant agricultural crops and native vegetation in California's Central Valley, over 40 different organic compounds were identified as being emitted from around 30 different species.[29] Low emitters included rice and wheat and the largest emitters were pistacio and tomato with total assigned plant emissions up to $30\,\mu g\,[g\,(dry\,wt)]^{-1}\,h^{-1}$ at temperatures around 30 °C. In a survey of several Californian tree species, emission rates, standardized to 30 °C, were as high as 37 and $49\,\mu g\,[g\,(dry\,wt)]^{-1}\,h^{-1}$ for Liquidambar and Carrotwood, respectively. Both species were predominantly isoprene emitters.[30]

Sitka spruce is the most abundant tree species in the UK, being the principal species over an area of about 530 kha or 25% of the total woodland area in the country. It emits both isoprene and monoterpenes.[13,14] Of the 21 most abundant grass and herbaceous species in the UK, only purple moor grass, bracken, and common gorse were found to emit isoprene, and only ivy and cocksfoot grass were found to emit a monoterpene.[26]

Effect of Temperature and Light Intensity on Emission Rates

Isoprene emission rates are strongly dependent on leaf temperature and, especially, light intensity. As leaf temperatures increase at a given light intensity, isoprene emissions increase exponentially, pass through a maximum at temperatures around 35 to 40 °C, and then decline, probably as a result of leaf damage and enzyme inactivation.[31] At a given temperature, isoprene emission rates increase with increasing light intensity up to a point of light saturation at around $800\,\mu mol\,m^{-2}\,s^{-1}$. Several different models attempting to describe changes in isoprene emission rates with changes in temperature, light (PAR, photosynthetically active radiation), and time of day have been developed. Isoprene emission rates are now believed to be under enzymatic control, linked to ATP levels which are dependent on light and temperature.

Monoterpene emission rates are primarily affected by leaf temperature, and there have been contradictory reports of a light dependency of monoterpene emission rates. Tingey *et al.*[32] reported no increase in monoterpene emissions from Slash pine (*Pinus elliottii*) under conditions of constant temperature when light intensity was increased from 0 to $800\,\mu mol\,m^{-2}\,s^{-1}$. However, Simon *et al.*[33] found that the average daytime emission rate of α- and β-pinene from maritime pine (*Pinus pinaster*) was around $0.1\,\mu g\,g^{-1}\,h^{-1}$, the average night-time emission rate $0.06–0.07\,\mu g\,g^{-1}\,h^{-1}$, and their emission rates were found to be a linear function of light intensity. Emissions from Sitka spruce were found to increase slightly with an increase in light intensity between 400 and $1000\,\mu mol\,m^{-2}\,s^{-1}$,[13] diurnal changes were observed in emission rates from Red

[29] A.M. Winer, J. Arey, R. Atkinson, S.M. Aschman, W.D. Long, C.L. Morrison, and D.M. Olszyk, *Atmos. Environ.*, 1992, **26A**, 2647.
[30] S.B. Corchnoy, J. Arey, and R. Atkinson, *Atmos. Environ.*, 1992, **26B**, 339.
[31] A.B. Guenther, R.K. Monson, and R. Fall, *J. Geophys. Res.*, 1991, **96**, 10799.
[32] D.T. Tingey, M. Manning, L.C. Grothaus, and W.F. Burns, *Plant Physiol.*, 1980, **20**, 797.
[33] V. Simon, B. Clement, M.L. Riba, and L. Torres, *J. Geophys. Res.*, 1994, **99**, 16501.

pine (*Pinus densiflora*) maintained at a constant temperature[34] and these findings were also supported by Steinbrecher *et al.*[35] in a study of Norwegian spruce (*Picea abies*). Thus, unlike isoprene whose emission is almost nil or very low during night-time, monoterpenes can still be emitted in the dark.

Other Factors Affecting Biogenic VOC Emissions

Vapour pressure and abundance of individual monoterpenes within the plant have been found to play a role in determining emissions. However, this is not universal for all monoterpenes. Lerdau *et al.*[36] found that needle concentrations of Δ^3-carene were not correlated to emission rates, whereas α- and β-pinene concentrations were. Yokouchi and Ambe[34] suggested that factors affecting the accumulation of the monoterpenes may contribute to the seasonal variability observed in plant tissue concentrations and emission rates.

There are differences in emissions at different stages of plant development. The emission of isoprene in very young leaves of velvet bean (*Mucuna* sp.) is negligible but increases over 100-fold before declining in older leaves. Studies of Eucalyptus at a field site in Portugal (authors' unpublished data) showed that for young plant tissues, emission rates of monoterpenes (approximately $7.5\,\mu g$ $[g\,(dry\,wt)]^{-1}\,h^{-1}$ under ambient conditions) were almost ten-fold greater than from old leaves and on average five-fold more isoprene was produced by young trees (approximately $63\,\mu g$ $[g\,(dry\,wt)]^{-1}\,h^{-1}$) than old. The proportion of different monoterpenes emitted also varied between different age groups. Guenther *et al.*[31] also considered leaf age as one of the many factors which can affect emission rate variability.

It was observed recently[37] that the emissions from common gorse (*Ulex europaeus*) in flower were predominently of isoprene ($>2\,\mu g$ $[g\,(dry\,wt)]^{-1}\,h^{-1}$ under ambient conditions), α-pinene ($0.4\,\mu g$ $[g\,(dry\,wt)]^{-1}\,h^{-1}$) and α-terpineol ($0.3\,\mu g$ $[g\,(dry\,wt)]^{-1}\,h^{-1}$), but when not in flower the same plant ceased to emit significant quantities of α-pinene and α-tepineol but emitted large quantities of linalool ($0.7\,\mu g$ $[g\,(dry\,wt)]^{-1}\,h^{-1}$). Emissions from Valencia orange groves were also found to be ten-fold higher during blossoming than at other times of the year, and linalool emission rates increased from around 0.1 to $13\,\mu g\,[g\,(dry\,wt)]^{-1}\,h^{-1}$.[28]

There have been several recent investigations of the effects of elevated CO_2 on isoprene emission rates. Preliminary work suggested a doubling of isoprene emission rate from oak after prolonged exposure to elevated CO_2 concentrations.[37] This is in agreement with previously published work on emissions from oak[38] but in contrast to findings by the same authors that emissions of isoprene from aspen were reduced after exposure to CO_2. Tingey *et al.*[39] reported an increase in

[34] Y. Yokouchi and Y. Ambe, *Plant. Physiol.*, 1984, **75**, 1009.

[35] R. Steinbrecher, W. Shurmann, R. Schonwitz, G. Eichstadter, and H. Ziegler, Proceedings of EUROTRAC Symposium 90, ed. P. Borrell *et al.*, SPB Academic Publishing, Den Haag, The Netherlands, 1991, p. 221.

[36] M. Lerdau, S. B. Dilts, H. Westberg, B. K. Lamb, and E. J. Allwine, *J. Geophys. Res.*, 1994, **99**, 16 609.

[37] S. C. Duckham, R. A. Street, C. Boissard, and C. N. Hewitt, EUROTRAC Annual Report 1993, 1994, p. 147.

[38] D. T. Sharkey, F. Loreto, and C. F. Delwiche, *Plant Cell Environ.*, 1991, **14**, 333.

isoprene emissions from *Quercus virginiana* (oak) with exposure to reduced CO_2 concentrations. Isoprene emissions were not affected by water stress in this study, but they have recently been associated with other plant stresses.

Changes in relative humidity are believed to have little if any effect on monoterpene emissions and a slight effect on isoprene emissions. Guenther *et al.*[31] observed a 2.4% increase in isoprene emissions from eucalyptus with every 10% increase in relative humidity. Monoterpenes appear to play a role as an anti-herbivore adaptation in some cases. Monoterpene concentrations were greatest in young tissues of camphorweed, making them more noxious to larvae.[40] Zinc stress in Japanese mint resulted in reduced monoterpene levels.[41] Several studies indicate that emissions of VOCs can be affected by wounding, both qualitatively and quantitatively. Large differences were found in emissions from excised and intact *Eucalyptus* leaf tissue, after 'improper handling' of Valencia orange branches[28] and after the mowing of grassland.[42]

Much further research is required to enhance understanding of the underlying mechanisms responsible for the observed variations in biogene VOC emission rates. Progress has been made in the study of isoprene emissions but this work must now be expanded to encapsulate the whole range of biogenic emissions, not only the monoterpenes, but also light hydrocarbons and oxygenated compounds. Only then will there be significant progress towards a wider understanding of how biogenic emissions affect the complexities of tropospheric chemistry.

4 Air Concentrations and Emission Fluxes

Emission Fluxes and Global Inventories

Emission rates and fluxes of biogenic hydrocarbons from different plant species have been measured at a variety of sites since the 1970s using different techniques,[7,11,13–16,34,43] and Table 1 and 2 show representative emission rates and fluxes from some of these investigations. In early measurements, isoprene and monoterpenes were found to be the dominant biogenic emissions, but more recently the emission of a huge range of VOCs has become apparent.

Emission inventories play an interactive role with atmospheric measurements and with model studies. The inventory can be used as an input condition for the model whose results are then validated by comparison with observations, and it can also be used as an input condition for the model whose results are used to estimate emission rates. Global estimates are often based on simple extrapolations and can have significant errors, likely to be a factor of three or more for annual averages and much more for specific time periods. Tables 3 and 4 summarize biogenic VOC emission estimates globally and for some specific countries.

For the continental land masses of the world, the main biogenic sources of

[39] D.T. Tingey, E. Rosemary, and M. Gumpertz, *Planta*, 1981, **152**, 565.

[40] C.A. Milhaliak and D.E. Lincoln, *J. Chem. Ecol.*, 1989, **15**, 1579.

[41] A. Misra, *Photosynthetica*, 1992, **26**, 225.

[42] G. König, M. Brunda, H. Puxbaum, C.N. Hewitt, S.C. Duckham, and J. Rudolph, *Atmos. Environ.*, 1995, **29**, 861.

[43] H. Knoppel, B. Versino, and A. Peil, Proceedings of 2nd European Symposium on the Physico-Chemical Behaviour of Atmospheric Pollutants, 1981.

Table 1 Emission rates of biogenic hydrocarbons from different plant species

Reference	Location	Time	Species	Compound	Emission rates (μg [g (dry wt)]$^{-1}$ h^{-1})
16	Washington, US	July afternoon	Coniferous	Total monoterpenes†	1.6
15	Florida, US	July afternoon	Coniferous	Total monoterpenes†	3.4
7	Leningrad, USSR	July afternoon	Scots pine (*Pinus sylvestris*)	Total monoterpenes†	10
43	Italy	July afternoon	Scots pine (*Pinus sylvestris*)	Total monoterpenes†	0.5
	Italy	July afternoon	Norwegian spruce (*Picea abies*)	Total monoterpenes†	0.2
34	Japan	July afternoon	Red pine (*Pinus densiflora*)	Total monoterpenes†	0.1
35	Germany	July afternoon	Norwegian spruce (*Picea abies*)	Total monoterpenes†	1.4
12	Central Sweden	July afternoon (averaged)	Scots pine (*Pinus sylvestris*)	Total monoterpenes†	0.7 ± 0.3
		July afternoon (averaged)	Norwegian spruce (*Picea abies*)	Total monoterpenes†	0.4 ± 0.6
		July maximum	*P. sylvestris*/*P. abies*	Total VOCs	0.6
		Winter minimum	*P. sylvestris*/*P. abies*	Total VOCs	0.2
		May ($0 < $ PAR $ < 1500$; $7 < T < 28$)	Scots pine (*Pinus sylvestris*)	Σ(Δ^3-carene, α-pinene)	0.02–1.39
		June ($10 < $ PAR $ < 1900$; $17 < T < 32$)		Σ(Δ^3-carene, α-pinene)	0.007–0.99
		July ($10 < $ PAR $ < ?$; $12 < T < ?$)		Σ(Δ^3-carene, α-pinene)	0.06–1.65
		Aug. ($0 < $ PAR $ < 1850$; $8 < T < 22$)		Σ(Δ^3-carene, α-pinene)	0.02–0.13
		Oct. ($280 < $ PAR $ < 850$; $9 < T < 11$)		Σ(Δ^3-carene, α-pinene)	0.58–2.14
		May ($220 < $ PAR $ < 1130$; $12 < T < 20$)	Norwegian spruce (*Picea abies*)	Σ(Δ^3-carene, α-pinene)	0.09–0.52
		June ($1190 < $ PAR $ < 1320$; $18 < T < 20$)		Σ(Δ^3-carene, α-pinene)	0.18–0.38
		July ($? < $ PAR $ < ?$; $16 < T < 19$)		Σ(Δ^3-carene, α-pinene)	0.01–2.91
		Aug. ($1050 < $ PAR $ < 1600$; $19 < T < 21$)		Σ(Δ^3-carene, α-pinene)	0.29–0.57
		Oct. ($210 < $ PAR $ < 950$; $19 < T < 21$)		Σ(Δ^3-carene, α-pinene)	0.09–0.83

No.	Location	Conditions	Species/Category	VOC	Value
12	Bavaria, Germany	July afternoon (averaged)	Norwegian spruce (*Picea abies*)	Total monoterpenes†	0.5
44	US	Annual emission factors*	High isoprene deciduous	Isoprene	13.6
			High isoprene deciduous	α-Pinene	0.06
			High isoprene deciduous	other monoterpenes	0.33
			High isoprene deciduous	other VOCs	1.82
			Low isoprene deciduous	Isoprene	5.95
			Low isoprene deciduous	α-Pinene	0.05
			Low isoprene deciduous	other monoterpenes	0.3
			Low isoprene deciduous	other VOCs	1.44
			Non-isoprene deciduous	Isoprene	0
			Non-isoprene deciduous	α-Pinene	0.07
			Non-isoprene deciduous	other monoterpenes	0.35
			Non-isoprene deciduous	other VOCs	1.54
			Non-isoprene coniferous	Isoprene	0
			Non-isoprene coniferous	α-Pinene	1.13
			Non-isoprene coniferous	other monoterpenes	1.78
			Non-isoprene coniferous	other VOCs	1.35
36	Cascade mountains, Oregon, US	June–September ($3\,^{\circ}C < T < 40\,^{\circ}C$)	Ponderosa pine (*Pinus ponderosa*)	Total monoterpenes (β-pinene, α-pinene, Δ^3-carene, limonene, β-phellandrene)	0.1–8.6
31	US	July afternoon⨝	49 US tree species	Isoprene	<0.1–70
	US	—	49 US tree species	Monoterpenes	<0.1–3
	US	—	49 US tree species	Other VOCs	<0.5–5

* In μgC $[\text{g (dry wt)}]^{-1}\,h^{-1}$, standardized at PAR $= 800\ \mu\text{mol}\,m^{-2}\,s^{-1}$ and $T = 30\,^{\circ}C$.

† Standardized at $T = 20\,^{\circ}C$.

⨝ Standardized at PAR $= 1000\ \mu\text{mol}\,m^{-2}\,s^{-1}$ and $T = 30\,^{\circ}C$.

VOCs are thought to be vegetation. The first emission inventory for plant emissions in the USA was made by Zimmerman[15] who divided the country into four regions to assess total isoprene and monoterpene emissions. Estimates of isoprene and monoterpene fluxes were 350 and 480 Tg yr^{-1} respectively. Rasmussen and Khalil,[49] using previous empirical measurements of the isoprene emission rates from a few species as the basis for inventories of several ecosystem types, extrapolated to a global estimate of 450 Tg yr^{-1} for isoprene.

The most authoritative estimate is that of Guenther *et al.*[48] who determined four categories of chemical species: isoprene, monoterpenes, other reactive VOCs (ORVOCs) and other VOCs (OVOCs). This model consists of a grid of resolution $0.5° \times 0.5°$ and generates hourly emission estimates. Temperature and light dependences of VOC emissions were used. The ocean is described as a function of geophysical variables from general circulation model and ocean colour satellite data. The annual isoprene flux is 420 TgC yr^{-1} and 130 TgC yr^{-1} for monoterpenes. Tropical drought-deciduous and savanna woods are predicted to contribute about half of all the global natural VOCs; other woodlands, croplands, and shrublands account for between 10 and 20%. The estimated isoprene emissions in temperate regions are a factor of three or more higher than previous estimates. Isoprene and monoterpenes remain the largest fraction. However, it is becoming increasingly clear that a variety of partially oxidized hydrocarbons, principally alcohols, are also emitted by vegetation and are currently unaccounted for adequately in these inventories.

Emission estimates of biogenic VOCs for regions in Europe are relatively uncertain and to date are largely dependent on US emission data. This increases the uncertainty in the estimates by an unknown amount, associated with emission rate factors and/or algorithms used for vegetation species that have not been measured. Janson[12] estimated the Swedish boreal forest monoterpene emissions to be 370 ± 180 Gg VOC yr^{-1}.

At the moment there is a paucity of high quality biogenic VOC emission data suitable for model input. Temporal resolution of existing inventories is almost uniformly poor and much remains to be done. However, a few global inventories have won acceptance by the modelling community, and although there are considerable uncertainties in these estimates, it is believed that the majority of global VOC emissions are from natural and not anthropogenic sources. For the USA and Canada, extensive field work has been done to validate the emission algorithms, leading to a reasonably high confidence in the inventories. But for the rest of the world, much effort is needed to improve inventories, especially for the tropical and boreal forests.

[44] B. Lamb, D. Gay, and H. Westberg, *Atmos. Environ.*, 1993, **11**, 1673.

[45] G. Enders *et al.*, *Atmos. Environ.*, 1991, **26A**, 171.

[46] J. Duyzer, TNO Report INW-R93/912, 1993.

[47] C. Geron, A. B. Guenther, and T. E. Pierce, *J. Geophys. Res.*, 1994, **99**, 12 773.

[48] A. Guenther, C. N. Hewitt, D. Erickson, R. Fall, C. Geron, T. Graedel, P. Harley, L. Klinger, M. Lerdau, W. A. McKay, T. Pierce, B. Scholes, R. Steinbrecher, R. Tallamraju, J. Taylor, and P. R. Zimmerman, *J. Geophys. Res.*, 1995, **100**, 8873.

[49] R. A. Rasmussen and K. A. K. I. Kalil, *J. Geophys. Res.*, 1988, **93**, 1417.

Table 2 Emission fluxes of biogenic hydrocarbons from different plant species

Ref.	Location	Time	Species	Compound	Model	Enclosure	Gradient	REA
						Emission fluxes (μg m^{-2} h^{-1})		
11	Washington State	Sept.–Nov. (9–24 °C)	Douglas fir (*Pseudotsuga*)	α-Pinene		46 (9–700)	230 (76–1320)	
45	Central Europe (Germany)	May/June—over 24 h	Norwegian spruce (*Picea abies*)	α-Pinene		8.1–51.5		
				β-Pinene		0.4–43.6		
		Sept.—over 24 h		α-Pinene		6.5–61.6		
				β-Pinene		0.32–5.8		
46	Speulderbos forest (Europe)	8th July over 24 h	Douglas fir (*Pseudotsuga*)	α-Pinene			36–216	
				β-Pinene			up to 288	
44	US	June max. average	Coniferous	Terpenes	900–1800			
	US	July max. average	Coniferous	Terpenes	940–1860			
47	Anderson County, Tennessee, US	All the year	Forested areas	Isoprene	4790*–7950†			
				Total BVOC	5650*–8810†			
	York County			Isoprene	6100*–10 200†			
				Total BVOC	6890*–11 000†			
	Choctaw County			Isoprene	4770*–7910†			
				Total BVOC	6260*–9400†			
	Emanual County			Isoprene	4440*–7300†			
				Total BVOC	5940*–8840†			
48	US	—	91 woodland-cover type	Total BVOC¥	800–11 000			
	US		Scrub woodlands	Total BVOC¥	800–4300			
	US		Deciduous/coniferous	Total BVOC¥	2200–11 000			
	US		Woodlands (average)	Total BVOC¥	5100			
33	France	June (day + light mean values)	Maritime pine (*Pinus pinaster*)	α-Pinene		63		
				β-Pinene		72		
				Myrcene		30		
				Δ^3-Carene		31		
				Limonene + cineole		29		
13	Rivox forest (SW Scotland)	July—Day averages (Extrapolation from enclosure emission rates)	Sitka spruce	Isoprene		658 ± 482 (70–1585)	318	13 ± 150
				α-Pinene		89 ± 30 (62–165)		241 ± 320
				β-Pinene		182 ± 118 (71–437)		—
				Myrcene		70 ± 20 (44–110)		30 ± 65
				Uncertain (^)		157 ± 81 (72–348)		415
				Limonene				
	Rivox forest (SW Scotland)	July—Night averages (enclosure emission rates)	Sitka spruce	Isoprene		33 ± 26 (3–77)	140 ± 130	360 ± 300
				α-Pinene		62 ± 1 (60–63)		
				β-Pinene		32 ± 6 (28–44)		
				Myrcene		46 ± 3 (42–50)		
				Uncertain (^)		73 ± 6 (65–81)		

* In μgC m^{-2} h^{-1}, standardized at PAR = 500 μmol m^{-2} s^{-1} and T = 30 °C; † in μgC m^{-2} h^{-1}, standardized at PAR = 1500 μmol m^{-2} s^{-1} and T = 30 °C; ¥ standardized at PAR = 1000 μmol m^{-2} s^{-1} and T = 30 °C; ^ β-thujene or β-phellandrene.

C.N. *Hewitt* et al.

Table 3 Global emission inventories of biogenic VOCs

Reference	Biogenic VOCs (TgC yr^{-1})	Comments
	Total VOCs	
50	432	
51	480	
53	500–825	
52	842–972	Uncertainties are order of three
	6–36	Ocean only
	6	Microbial production
54	491	
	294	Northern hemisphere
	197	Southern hemisphere
	360	Tropical ecosystems
	132	Non-tropical ecosystems
48	1100	Estimate for 1990
	750	Total woods
	120	Crops
	160	Shrub
	4	Ocean
	72	Others
	Isoprene	
15	350	
49	452	
55	450	
52	350–450*	
56	285	
48	420	Estimate for 1990
54†	250	
	132	Northern hemisphere
	117	Southern hemisphere
	208	Tropical ecosystems
	41	Non-tropical ecosystems
	Monoterpenes	
15	480	
48	130	Estimate for 1990
54	147	
	105	Northern hemisphere
	42	Southern hemisphere
	81	Tropical ecosystems
	66	Non-tropical ecosystems
	Other reactive VOCs	
52	480	
48	280	Estimate for 1990

* Extrapolation from US inventories would suggest that these sources may be overestimated.
† By scaling up US emission rates to the global level using a net primary productivity ratio.

[50] R. A. Rasmussen and F. Went, *Proc. Natl. Acad. Sci. USA*, 1965, **53**, 215.
[51] E. Robinson and R. Robbins, Final Report PR-6756, Standford Research Institute, Menlo Park, California, 1968.
[52] H. B. Singh and P. R. Zimmerman, in 'Gaseous Pollutants: Characterization and Cycling', ed. J. O. Nriagu, John Wiley & Sons, New York, 1992.
[53] F. Fehsenfeld, J. Calvert, R. Fall, P. Goldan, A. B. Guenther, C. N. Hewitt, B. Lamb, S. Liu, M. Trainer, H. Westberg, and P. Zimmerman, *Global Biogeochem. Cycles*, 1992, **6**, 389.
[54] J.-F. Müller, *J. Geophys. Res.*, 1992, **97**, 3787.

Table 4 Biogenic VOCs inventories for different countries (kTC yr⁻¹)*

References	Isoprene	†	Monoterpenes	†	Reactive VOCs	†
N. Australia						
57	25 000	14 800				
Germany						
58					349	960
59					970	960
60	120	310			193	
Hungary						
58					22	230
59					70	230
61					156	230
60	82	99				
Spain						
58					1132	1400
59					6600	1400
60	138	650				
Sweden						
15			800–1200	340		
58					930	910
59					600	910
12			370	340		
60	108	300				
United Kingdom						
58					38	280
59					50	280
62			71	120		
63 (estimate for 1985)	2–62	84				
64		84	167	120	211	280
65 (estimate for 1989)	2–36	84				
60	23	84				
United States						
15	15 000	21 000	50 000	8000		
66	4900	21 000			30 700	42 000
52	7000–20 000	21 000			25 000–50 000	42 000
44	3000–14 700	21 000	8300–33 300	8000	22 800–80 100	42 000

† Estimated from model (reference 48).
* kTC yr⁻¹: kiloton carbon per year.

[55] J. Dignon and J. A. Logan, *Eos Trans. AGU*, 1990, **71**, 1260.
[56] D. Turner, J. Baglio, D. Pross, A. Wones, B. McVeety, R. Vong, and D. Phillips, *Chemosphere*, 1991, **23**, 37.
[57] G. P. Ayers and R. W. Gillet, *J. Atmos. Chem.*, 1988, **7**, 177.
[58] B. Lubkert and W. Shoepp, IIASA working paper WP-89-082, 1989.
[59] V. Adryukov and A. Timofeev, 4th ECE task force on volatile organic compounds, Schwetzingen, Germany, 1989.
[60] D. Simpson, A. Guenther, C. N. Hewitt, and R. Steinbrecher, *J. Geophys. Res.*, in the press.
[61] A. Molnar, *Atmos. Environ.*, 1990, **25A**, 2855.
[62] OECD, Environment Monograph 21, OECD Map Emission Inventory, Paris, 1989.
[63] D. Simpson and O. Hov, EMEP MSC-W note 2/90, 1990.
[64] C. Anastasi, L. Hopkinson, and V. Simpson, *Atmos. Environ.*, 1991, **25A**, 1403.
[65] D. Simpson, EMEP MSC-W note 2/91, 1991.
[66] B. Lamb, A. Guenther, D. Gay, and H. Westberg, *Atmos. Environ.*, 1987, **21**, 1695.

Air Concentration Measurements

Measurements of air concentrations of isoprene and monoterpenes and their diurnal variations have been made since the late 1970s,[67–71] and some typical data are summarized in Table 5. Different diurnal patterns have been observed for isoprene and monoterpenes. Generally, isoprene concentrations increase sharply in the early morning after sunrise with a maximum in the afternoon, while the air concentrations of monoterpenes (*e.g.* α-pinene) during the diurnal cycle are the inverse of that observed for isoprene. This is due to the fact that emission of isoprene from vegetation, depending on both temperature and PAR, is almost nil during the night, while monoterpenes are still emitted during the night-time, since their emissions depend mainly on temperature, and are not very sensitive to PAR. In a polluted area, isoprene and monoterpenes can be destroyed by reactions with O_3 and $^\bullet OH$ radicals during the day, and during the night they can also react with $NO_3^\bullet$ radicals, in addition to O_3. Since some monoterpenes, especially α-pinene, can react more quickly with O_3 and $NO_3^\bullet$ than does isoprene, their concentrations during the day may reach a minimum even though their emissions are at maximum. However, this will be much less important if the concentrations of O_3, $^\bullet OH$, and $NO_3^\bullet$ radicals at the sampling site are very low.

In clean air, dispersion rather than chemistry is the most important factor determining air concentrations of the monoterpenes. The atmosphere during the night-time is much more stable than during the daytime, the wind speed dropping significantly during the night. Nocturnal temperature inversions have also been frequently observed to influence ambient biogenic VOC concentrations. The capping effect of this inversion tends to trap the monoterpenes which are still emitted during the night, and this results in their concentrations increasing until the morning breakdown of this inversion leads to rapid mixing and dilution.[53] However, a clear diurnal pattern of hydrocarbon concentrations may be observed only during fine days, while their concentrations during wet and/or very windy days may show a large and unsystematic variation.[69]

In a recent study, diurnal air concentrations of isoprene and monoterpenes above the canopy of a Sitka spruce forest (Scotland) where the weather was cold and wet and ozone concentrations were low, were made for two consecutive days.[14] The isoprene concentrations showed a clear diurnal pattern [Fig. 1(a)], concentrations being a minimum during the night when temperatures and PAR were low, and reached a maximum during or after midday when temperatures and PAR were higher. For the monoterpenes [*e.g.* α-pinene, Fig. 1(b)], the diurnal concentrations were rather unsystematic, reached a maximum around and after midnight, and a minimum during the day for the first sampling day, but the other way around for the second day. The average concentrations of these compounds, the total concentrations of monoterpenes and their relative compositions during the day and night (shown in Table 6) also vary. For

[67] M. W. Holdren, H. H. Westberg, and P. R. Zimmerman, *J. Geophys. Res.*, 1979, **84**, 5083.

[68] R. J. B. Peters, J. A. D. V. R. V. Duivenbode, J. H. Duyzer, and H. L. M. Verhagen, *Atmos. Environ.*, 1994, **28**, 2413.

[69] Y. Yokouchi and Y. Ambe, *J. Geophys. Res.*, 1988, **93**, 3751.

[70] J. M. Roberts, F. C. Fehsenfeld, D. L. Albritton, and R. E. Sievers, *J. Geophys. Res.*, 1983, **88**, 10 667.

[71] R. W. Janson, *J. Atmos. Chem.*, 1992, **14**, 385.

Table 5 Typical air concentrations (ppbv) of isoprene and monoterpenes from selected terrestrial sites

Compound	Georgia[75] Forest S	Niwot Ridge[70] 3 km Rocky Mountains S	A	Italy[72] Rome Forest S	France[73] Rural S	Appalachian[74] Mountains Forest A
Isoprene	1.4	0.63	0.11		0.19	<0.1–1.1
α-Pinene	0.8	0.05–0.14	0.07	1.5		0.03–0.05
β-Pinene	0.43	0.97–0.08	0.07	0.18		
Δ³-Carene	0.9	0.05		0.06		
Camphene	0.09	0.04				
Sabinene						
D-Limonene	0.08	0.03–0.05	0.05	0.04		
β-Myrcene	0.068					
p-Cymene						

Compound	Brazil[76] Amazon basin Forest S	A	South eastern US[53] Rural S (04:00)	S (16:00)	Southwest Scotland[14] Sitka spruce Forest AD	AN
Isoprene	2.04	5.45	0.9	6.3	0.59	0.302
α-Pinene	0.1	0.2	0.75	0.3	0.027	0.025
β-Pinene	0.03	0.01	0.4	0.17	0.012	0.015
Δ³-Carene	0.01	<0.01				
Camphene	0.03	0.04	0.062	0.012	0.006	0.006
Sabinene					0.008	0.008
D-Limonene					0.016	0.015
β-Myrcene					0.021	0.027
p-Cymene			0.15	0.008		

S: summer; A: autumn; AD: autumn daytime; AN = autumn night-time.

isoprene, concentrations during the second day were higher than during the first day, and the daytime concentrations were all higher than at night-time. The daytime concentrations for monoterpenes during the first day were generally lower than the night-time concentrations. During the second day, however, the daytime concentrations are generally larger. The most abundant monoterpene during the day was α-pinene/limonene for the first day and α-pinene for the second day; and during the night, α-pinene for the first day and β-myrcene for the second day.

Data on air concentrations of hydrocarbons at different heights above the canopy of the forest, which are necessary for the calculation of emission fluxes by the gradient method, are very limited. Peters *et al.*[68] measured concentrations of the monoterpenes at several different heights above a pine forest, and a concentration gradient could be found for some monoterpenes. In the recent

[72] P. Ciccioli, E. Brancaleoni, M. Possanzini, A. Brachetti, and C. Di Palo, Proceedings of the Physico-Chemical Behaviour of Atmospheric Pollutants, Varese, Italy, 1984.

[73] M. Kanakidou, B. Bonsang, J. C. Le Roulley, G. Lambert, D. Martin, and G. Sennequier, *Atmos. Environ.*, 1988, **23**, 921.

[74] R. L. Seila, in 'Impact of Natural Emissions', ed. V. Aneja, Air Pollution Control Association, Pittsburgh, Pennsylvania, 1984, p. 125.

[75] R. W. J. Shaw, A. L. Crittende, R. K. Stevens, D. R. Cronn, and V. S. Titov, *Environ. Sci. Technol.*, 1983, **17**, 466.

[76] P. R. Zimmerman, J. P. Greenberg, and C. Westberg, *J. Geophys. Res.*, 1988, **93**, 1407.

Table 6 Mean concentrations of isoprene and monoterpenes and relative compositions of monoterpenes above Sitka spruce forest, at two different heights (taken from C. N. Hewitt et al.[14])

Compound	Concentration (ppbv)				Relative composition (%)			
	17.5 m		22.5 m		17.5 m		22.5 m	
	day	night	day	night	day	night	day	night
21 Aug. 1994								
Isoprene	0.446	0.244	0.379	0.217				
α-Pinene	0.026	0.043	0.014	0.032	26.75	33.2	23.22	30.59
Camphene	0.006	0.009	0.004	0.008	6.69	6.87	6.3	7.2
Sabinene	0.010	0.010	0.005	0.008	9.93	7.49	8.13	7.2
β-Pinene	0.016	0.027	0.009	0.020	16.41	21.16	14.93	19.03
β-Myrcene	0.015	0.014	0.012	0.022	15.78	10.42	20.56	20.83
D-Limonene	0.023	0.027	0.016	0.016	24.45	20.85	26.87	15.15
Total (monoterpene)	0.096	0.130	0.060	0.106				
22 Aug. 1994								
Isoprene	0.850	0.367	0.699	0.378				
α-Pinene	0.031	0.015	0.036	0.010	28.1	17.91	38.24	15.02
Camphene	0.007	0.004	0.006	0.004	6.39	5.28	6.77	6.26
Sabinene	0.011	0.007	0.005	0.008	9.95	8.22	8.16	9.86
β-Pinene	0.012	0.008	0.010	0.005	11.13	9.57	10.53	8.14
β-Myrcene	0.030	0.035	0.026	0.035	27.37	42.45	28.36	54.46
D-Limonene	0.019	0.014	0.007	0.004	17.06	16.56	7.95	6.26
Total (monoterpene)	0.110	0.082	0.090	0.065				

Figure 1 Diurnal variation of hydrocarbon concentrations above Rivox forest, Scotland, at different heights during 21–22 August 1994. (a) Isoprene; (b) α-Pinene. (Taken from C. N. Hewitt *et al.*[14])

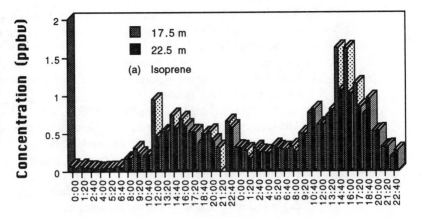

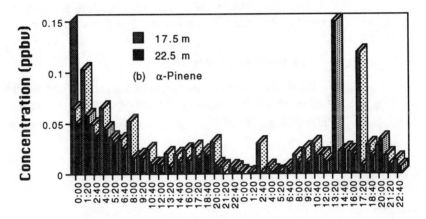

work of Hewitt *et al.*,[14] diurnal concentrations of isoprene and monoterpenes were measured at two different heights above a Sitka spruce forest. As can be seen in Table 6, the average concentrations of isoprene and monoterpenes at the lower height are generally greater than those at the higher level. The diurnal concentrations for isoprene and α-pinene at these two heights, as shown in Figure 1, show that most of the concentration gradients are positive (that is, the concentration at the lower height is larger than at the higher level), but negative gradients were also observed. This may indicate that negative fluxes calculated from the gradient method may occur, due to the complex diurnal variations of meteorological conditions.

5 Influence on Atmospheric Chemistry

Biogenic hydrocarbons are actively involved in the chemistry of the lower atmosphere, and indeed they may play a dominant role in certain aspects of the chemical behaviour of the atmosphere. The consequences of their emissions may be briefly summarized as follows:

(i) They react rapidly with ozone and hydroxyl radicals, forming, among other products, CO, and thereby impact directly on the oxidizing capacity of the troposphere.

(ii) In areas of high NO_x concentrations, they may contribute to the formation of ozone on regional scales and hence influence photochemical oxidant formation processes.

(iii) They may significantly contribute to the global carbon budget.

(iv) They may produce organic acids and hence contribute to the deposition of acidity in rural and remote continental areas.

(v) They are involved in the generation of organic nitrates which can sequester NO_x and facilitate its transport over large distances.

(vi) They can lead to organic aerosol formation and contribute to the degradation of visibility on the regional scale.

A thorough review of these issues is available in Fehsenfeld *et al.*[53] and they are covered in detail in Chapters 1 and 5 of this volume (pages 1 and 65).

6 Acknowledgements

We thank the Natural Environment Research Council and EC Environment Programme for funding, and Rachel Street, Alex Guenther, and Ray Fall for stimulating discussions.

The UK Hydrocarbon Monitoring Network

G. J. DOLLARD, T. J. DAVIES, B. M. R. JONES,
P. D. NASON, J. CHANDLER, P. DUMITREAN,
M. DELANEY, D. WATKINS, AND R. A. FIELD

1 Introduction

The importance of volatile organic compounds (VOCs) in the atmosphere is apparent when their role in atmospheric chemistry and their potential impact on human health are considered. The sources, distribution, and fates of VOCs are covered elsewhere in this volume (Chapter 1, page 1). This article will focus on recent developments in the United Kingdom relating to the measurement of VOC species.

VOCs form a broad grouping of compounds that have important impacts in terms of stratospheric ozone depletion, the formation of photochemical ozone, human health, and persistence in biotic and abiotic components of ecosystems. Within the UK, monitoring has focused mainly on the measurement of VOCs involved in the formation of ozone and those which are of direct consequence to human health. The discussion and presentations in this chapter are based upon the measurements made in support of the air pollution research programme of the UK Department of Environment. The measurements described are those for hydrocarbon species in the range C_2 to C_8 and range from simple grab samples to fully automatic network-derived data.

The data recorded as part of this programme provide information for several needs. As well as underpinning scientific research on atmospheric chemistry and facilitating development and validation of models, the data can be used to assess emissions inventories and to assess the impact of certain compounds, *e.g.* benzene, on human health. In addition, the member states of the United Nations Economic Commission for Europe have agreed a VOC protocol to the International Convention on Long Range Transboundary Air Pollution.[1] Under this protocol, emissions of VOCs are set to be reduced by 30% by the year 1999, based upon 1988 emissions data. All low molecular weight hydrocarbons are VOCs so they come into the scope of the VOC protocol; measurement data provide an important means of assessing the impact of such protocols.

[1] United Nations Economic Commission for Europe, 'Protocol to the 1979 Convention on Long Range Transboundary Air Pollution Concerning the Controls of Emissions of Volatile Organic Compounds or their Transboundary Fluxes', Report ECE/EB AIR/30, United Nations, Geneva, Switzerland, 1991.

There is also concern for the potential impact of hydrocarbon compounds on human health. Two compounds in particular, benzene and 1,3-butadiene, have been the subject of much discussion within the UK and air quality standards have been recommended for these compounds.[2,3] As a consequence, current hydrocarbon monitoring within the UK is geared to providing information on benzene and 1,3-butadiene concentrations to the public.

2 The United Kingdom Hydrocarbon Network

Following the publication of the UK Government White Paper on the environment[4] the Department of Environment set up its Enhanced Urban Monitoring Initiative. This was to establish comprehensive monitoring sites in city locations; the second stage of this initiative was the Hydrocarbon Network. The details of the initiative are given elsewhere.[5]

As a precursor to establishment of the network two pilot sites were established, one at a roadside site on Exhibition Road (Imperial College, London University) and the second at an urban background location in Middlesbrough. Ambient measurements have been made periodically at Exhibition Road since 1979. Automatic measurements were made for about 1 year during 1991/92. The Exhibition Road site was re-established at Gordon Square (University College London, London University) in 1993.

Measurements at Exhibition Road, London

The sampling strategy at Exhibition Road moved from hourly daytime values in 1979 to continuous hourly measurements in 1991/92. Concurrent with the increased sampling frequency the range of measured pollutants increased. The 1991/92 data were collected using a VOCAIR system for 26 hydrocarbons. The data from this period have facilitated assessments of the trends of pollutant concentrations at the site, the dominant pollutant emission sources at this site, and the accuracy of the UK VOC emission inventory.[6-8]

Since different sampling regimes and methodology were used at Exhibition Road it is only possible to carry out simple comparisons between these historical data for a limited number of compounds. Figure 1 summarizes data for aromatic compounds. Comparing daytime summer averages from 1979 through 1982/83

[2] Expert Panel on Air Quality Standards, 'Benzene', HMSO, London, 1994.

[3] Expert Panel on Air Quality Standards, '1,3-Butadiene', HMSO, London, 1994.

[4] 'This Common Inheritance', Government White Paper, HMSO, London, 1990.

[5] Quality of Urban Air Review Group, First Report: 'Urban Air Quality in the United Kingdom', Department of the Environment, London, 1993.

[6] Photochemical Oxidants Review Group (PORG), 'Ozone in the United Kingdom 1993,' Department of the Environment, London, 1993.

[7] R.G. Derwent, D. R. Middleton, R. A. Field, M. E. Goldstone, J. N. Lester, and R. Perry, 'Analysis and Interpretation of Air Quality Data from an Urban Roadside Location in Central London over the Period from July 1991 to July 1992,' *Atmos. Environ.*, 1995, **29**, 923.

[8] R.A. Field, M. E. Goldstone, J. N. Lester, and R. Perry, 'The Variation of Volatile Organic Compound Concentrations in Central London during the Period July 1991 to September 1992,' *Environ. Technol.*, 1994, **15**, 931.

Figure 1 Mean daytime concentrations of aromatic hydrocarbons for summertime (April–September) periods, calculated from various air quality studies since 1979

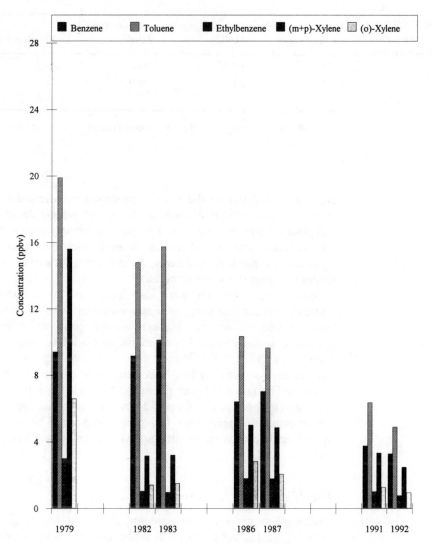

and 1986/87 to 1991/92 there is an apparent decline in ambient concentrations. This decline was thought to be primarily due to the impact of emission control technology and improved engine technology during the period on motor vehicle exhaust emissions.[9] The influence of meteorology on the ambient concentrations of the hydrocarbons measured at this site has also been demonstrated. Meteorology was highlighted as a particularly important influence during air pollution episodes, particularly during December 1987 and December 1991.[7,9]

Motor vehicle exhaust emissions (petrol and diesel), motor vehicle evaporative emissions and natural gas leakage emissions were identified as the predominant source of most of the hydrocarbons measured at this site. The dominance of motor vehicle emissions was not surprising given the roadside location of the site.

[9] R. A. Field, J. R. Brown, M. E. Goldstone, J. N. Lester, and R. Perry, 'Trends in Motor Vehicle Related Air Pollutants in Central London,' *Environ. Monit. Assess.*, 1995, in the press.

Table 1 Estimated* hydrocarbon speciation by carbon number for natural gas

Component	British Gas[10] (1992)	CONCAWE[11] (1986)	Reference 8	QUARG[5] (1993)
C2	68.5	70.0	66.4	68.7
C3	21.2	20.0	21.8	18.1
C4	10.3	10.0	11.8	9.4

* Calculated assuming the only carbon contribution is from ethane, propane, isobutane, and n-butane.

Data generated during the 1991/92 programme indicated high stability of the relative concentrations of most of the motor vehicle derived VOCs. Notable exceptions to this behaviour were propene which appeared to have a background source, and the butane and pentane isomers which are expected to have the most significant evaporative emissions; the concentrations of these compounds were elevated during the summer period.[8]

Natural gas emissions were also important for a number of species, in particular ethane and propane. Emission inventory estimates for motor vehicle emission and natural gas leakage showed good agreement with measured ambient concentrations.[7] The estimated speciation of natural gas at this site has compared well to other estimates: this comparison is given in Table 1.[8] A natural gas leakage occurred for two hours during August 1991 when gas mains were repaired. The ratios between measured hydrocarbons during these hours were compared to those prior to and after the natural gas leakage. Assuming all other source emission fingerprints (including background natural gas leakage) were stable, the difference can be attributed to the emission fingerprint of natural gas leakage.

The Hydrocarbon Network Structure

Hydrocarbon measurement sites are shown in Figure 2. The first site was commissioned in June 1991, and at the time of writing ten sites are operational including a non-network rural site located at Harwell, (Oxfordshire, S.E. England). The final two urban sites, at Liverpool and Southampton, will become operational during the summer of 1995. The monitoring sites, with the exception of Harwell, are located in urban or suburban population centres. In the context of the impact of the carcinogenic species the measurements are representative of exposure to the general urban population. The Harwell site is at a rural location and is used for development purposes as well as monitoring. The London 1 site is operated as a roadside site (4 m from the kerbside); all the other sites are at least 50 m from busy roadways.

AEA Technology, based at Culham, Oxfordshire, manage the network with local operator support for routine site operations. An independent QA/QC

[10] British Gas, personal communication.
[11] CONCAWE, 'Volatile Organic Compounds Emissions: An Inventory for Western Europe', Report no. 286, The Hague, The Netherlands, 1986.

Figure 2 UK hydrocarbon network sites

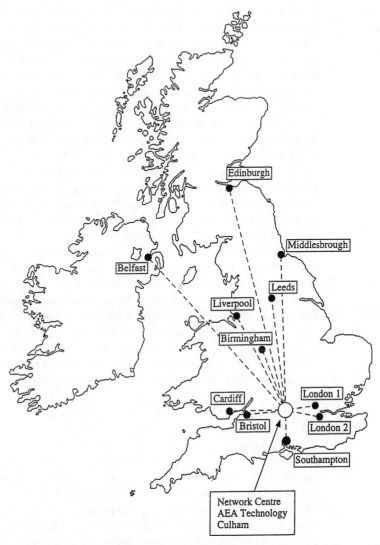

facility is provided by the National Physical Laboratory, Teddington, UK, which also supplies calibration gas mixtures for use at the sites.

Analytical Procedures

The equipment used to monitor hydrocarbons is the Chrompack VOCAIR analyser. The equipment allows continuous gas chromatographic determinations at hourly intervals. The system consists of an automatic thermo-desorption/cryogenic trapping system (Auto TCT), connected to a Chrompack gas chromatograph.

All sites have similar arrangements to facilitate sampling of ambient air. Air is drawn through a 110 mm diameter pipe by a high volume flow centrifugal pump

($5\,m^3\,minute^{-1}$). Air is drawn at a rate of $40\,cm^3\,min^{-1}$ from the high volume flow line through a Nafion drier. During the collection phase $10\,cm^3\,min^{-1}$ of the dried air is taken into the Auto TCT. The hydrocarbons are collected on a three-stage adsorption trap consisting of a glass tube (15 cm long, 0.6 cm diameter) packed with carbosieve, carbotrap, and carbosieve C held at $-20\,°C$ over a period of 35 minutes, giving a total air volume of $350\,cm^3$. Following the sample collection, the adsorption trap is back-flushed with helium and heated to $250\,°C$ to thermally desorb the hydrocarbons. The desorbed hydrocarbons are collected on a cryofocusing capillary trap consisting of a length of 0.53 mm internal diameter fused silica tube coated with Poraplot U (Chrompack). The trap is held at $-100\,°C$ during the 5 minute desorption phase. Following this phase the cryofocusing trap is heated rapidly to $120\,°C$ and the hydrocarbons injected onto the analytical column (PLOT fused silica, 25 m × 0.32 mm internal diameter, Al_2O_3 capillary column) The column is temperature-programmed from $50\,°C$ to $200\,°C$ in 3 stages, starting at $50\,°C$: $5\,°C\,min^{-1}$ for 5 min; $10\,°C\,min^{-1}$ for 5 min; $15\,°C\,min^{-1}$ for 5 min; and then held at $200\,°C$. The separated compounds are determined using a flame ionization detector (FID). Helium is used as the carrier gas. The VOCAIR GC controls the sampling and analytical parameters and facilitates one determination per hour. The integrated data are generated approximately two hours after the start of the collection phase.

Cooling requirements are met with liquid nitrogen which is held in pressurized Dewar flasks at each site. The rate of consumption of liquid nitrogen varies from site to site depending upon *inter alia* the length and exposure of pipework. As an average each VOCAIR system consumes about 50 litres of liquid nitrogen per day.

Data Handling

The data handling procedures are illustrated schematically in Figure 3. Each of the PCs installed alongside the VOCAIR equipment operates several software packages: PC anywhere, modem control software, and PCI chromatographic acquisition and analysis software. The PCI software collects the raw chromatogram (binary file), integrates the peaks, and produces a report file. The site PC is polled each hour via modem and all files are retrieved for storage on a workstation at Culham in the network management centre. A key species file is created on the site PC which extracts benzene and 1,3-butadiene data for rapid transmission and display on television teletext pages. Data on these compounds are made available after a two hour delay.

It is clear that the routine operation of the network generates large numbers of hourly data (requiring processing of many raw data files). Each site can generate in excess of 8000 hourly values per year. The data for all compounds are also made available for an on-line display of concentrations and retention times for data dissemination and QA/QC purposes. The compounds routinely reported are shown in Table 2; these represent a mixture of sources from natural gas, motor vehicle usage, and industrial activities.

Identification of the compounds is carried out using pattern recognition software (AEA MatchFinder). Monthly batches of the files integrated using PCI on the site PCs are processed using the pattern recognition software on the workstation. The output data are checked using commercial spreadsheet

Figure 3 Data handling and processing scheme for the hydrocarbon network

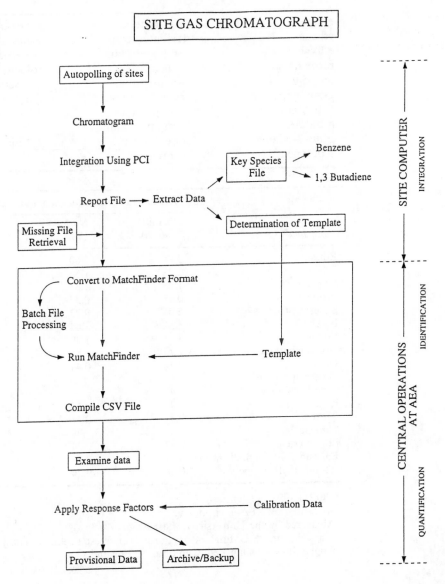

software. Final quantification is through application of response factors derived from the injection of certified standards at each site. Final ratification of the data is performed at the QA/QC unit.

3 Data Summaries and Discussion

Motor vehicle use contributes substantially to hydrocarbon emissions;[6] it is expected that patterns associated with motor vehicle emissions will be observed at all sites. Some or all of the sites may be influenced by other sources of hydrocarbons, *e.g.* industrial or petrochemical processes. This article concentrates on provisional data from the hydrocarbon network for the period

G.J. *Dollard* et al.

Table 2 Compounds reported from the hydrocarbon network

ethane	*cis*-2-butene	n-hexane
ethene	1,3-butadiene	n-heptane
ethyne	2-methylbutane	benzene
propane	(isopentane)	toluene
propene	n-pentane	(methylbenzene)
2-methylpropane	*trans*-2-pentene	ethylbenzene
(isobutane)	*cis*-2-pentene	1,3- and 1,4-dimethyl-
n-butane	2-methylpentane	benzene (*m*- + *p*-xylene)
trans-2-butene	3-methylpentane	1,2-dimethylbenzene
1-butene	isoprene	(*o*-xylene)

Table 3 Hydrocarbon concentrations at rural UK sites[1]

	Average concentrations (ppbv) for indicated period		
Compound	Harwell[2] [†] 1986–90	Norfolk[3] [†] 1989–91	Cumbria[3] [†] 1989–91
Ethane	3.95	2.80	2.10
Ethene	2.23	1.55	0.99
Propane	1.76	1.31	0.86
Propene	0.43	0.27	0.16
2-methylpropane	2.04	0.72	0.34
Butane	3.51	1.64	0.71
Ethyne	2.43	1.40	0.78
2-methylbutane	1.73	0.86	0.32
Pentene	0.74	0.44	0.19
2-methylpentane	0.61		
3-methylpentane	0.49		
Hexane	0.92	0.14	0.01
Benzene	0.81	0.73	0.34
Toluene	1.50	0.99	0.50
Ethylbenzene	0.37	0.24	0.13
1,3- and 1,4-dimethylbenzene	0.77	0.54	0.28
1,2-dimethylbenzene	0.38	0.24	0.15

[1] PORG 1993 (reference 6).
[2] Measured by AEA Technology.
[3] Measured by the University of East Anglia.
[†] Harwell, rural S.E. England (Oxfordshire); West Beckham, coastal site, Norfolk, E. England; Great Dun Fell, Cumbria, N. England.

January–December 1994; features of the data are discussed in broad terms. Some preliminary statistical analyses have been reported elsewhere,[12] as have analyses on similar UK data sets.[7]

Measurement of rural concentrations have been made at a few sites for many years. These measurements have been limited to grab samples at daily or weekly intervals followed by laboratory analysis on conventional gas chromatographic equipment. This limits the number of measurements that can be made. Table 3

[12] J. Derwent, P. Dumitrean, J. Chandler, T. J. Davies, R. G. Derwent, G. J. Dollard, M. Delaney, B. M. R. Jones, and P. D. Nason, 'A Preliminary Analysis of Hydrocarbon Monitoring Data from an Urban Site', Report AEA CS 18358030/005/2, AEA Technology, E5 Culham, OX14 3DB, UK, 1994.

lists the results of measurements on grab samples collected at three rural sites. The results indicate that there is a gradient between these sites. The Harwell site exhibits the largest influence by local sources and the Cumbrian site at Great Dun Fell exhibits the least influence and hence the lowest values. For comparison, data collected at remote Atlantic sites for benzene provide baseline values of 0.02 ppbv and 0.07 ppbv for the Southern and Northern hemispheres respectively.[13]

Data from the hydrocarbon network are summarized in Tables 4 and 5. Table 4 lists monthly statistics for all compounds for one site, Belfast. Table 5 gives annual statistics for all compounds at all sites.

The data for Belfast are in many ways typical of the other network sites; the monthly data reflect seasonal trends in concentration generated as a consequence of changes in emissions, meteorological factors, and removal from the atmosphere due to reaction with hydroxyl free radicals. The peak concentrations during episodes can be seen from the maximum hourly values given in Table 4; *e.g.* benzene and toluene data contain values of 26 and 60 ppb respectively during December 1994.

Belfast is unusual with respect to ethane and propane concentrations. The ethane concentrations observed are consistently lower than at other urban sites, whereas the propane concentrations are consistently higher. An explanation for this is probably that Belfast does not have supplies of natural gas and LPG systems are used for space heating.

It is interesting to note that several compounds show high maximum values; for example, ethene levels can be high at all sites. This particular compound is of interest due to its potential impact on vegetation. Isoprene is detected consistently and may be shown to be strongly correlated with ethyne and other motor-vehicle derived components. This is consistent with other observations,[7] and suggests a motor vehicle contribution to isoprene emissions, formerly solely attributed to vegetation sources.

Daily patterns can be clearly observed in the hydrocarbon data. The patterns are influenced by the nature of the source and meteorology or the degree of atmospheric mixing. Patterns are generally consistent across the sites with one or two exceptions. Figure 4 illustrates data for ethane at Belfast and Birmingham. Figure 5 illustrates data for benzene at Belfast and Birmingham. These plots summarize for each hour of the day the average of all the determinations made in that hour throughout the year.

Ethane is primarily derived from the leakage of natural gas. At Birmingham this shows the very typical behaviour of a pollutant with a constant surface source and is in marked contrast to that shown for benzene which is primarily from motor vehicle use. For a 24 hour period the ethane concentration is little affected by atmospheric chemistry (removal by reaction with ˙OH is relatively slow) and the plot for ethane illustrates the impact of atmospheric mixing and hence dilution on the ethane concentration. Ethane data for Belfast clearly illustrate the lack of the main ethane source, leakage of natural gas. The concentrations remain steady at a value substantially lower than that measured at rural sites (Table 3). It is likely that the rural site data are biased towards higher values as grab samples

[13] B. M. R. Jones, personal communication, 1995.

Table 4 Monthly statistics for all compounds at Belfast, January–December 1994 (ppbv)

Compound	January average	January hourly max.	February average	February hourly max.	March average	March hourly max.	April average	April hourly max.	May average	May hourly max.	June average	June hourly max.
ethane	1.97	8.42	2.71	12.17	1.51	3.66	1.40	5.81	1.26	5.42	0.73	2.52
ethene	2.12	25.47	2.85	38.11	1.00	5.29	1.00	12.62	1.12	11.66	0.83	7.42
propane	4.75	42.57	6.50	77.81	2.75	19.46	3.15	32.75	4.57	35.10	2.13	42.64
propene	1.28	15.81	1.63	25.92	0.89	3.42	1.02	8.36	1.25	7.51	1.09	5.05
ethyne	3.07	50.26	4.31	60.08	1.70	11.10	1.76	20.69	1.81	16.07	1.30	14.51
2-methylpropane	1.53	23.72	2.05	31.17	1.17	14.24	1.30	13.25	1.49	12.26	1.23	39.31
n-butane	3.32	53.72	4.27	67.84	2.10	16.83	2.23	22.26	2.77	17.11	2.36	60.17
trans-2-butene	0.19	2.74	0.24	3.74	0.18	0.71	0.24	1.61	0.28	4.25	0.28	2.78
1-butene	0.21	3.11	0.26	4.64	0.12	0.58	0.15	1.71	0.16	3.25	0.13	2.30
cis-2-butene	0.13	2.37	0.17	3.23	0.10	0.60	0.14	1.30	0.14	3.53	0.14	2.22
2-methylbutane	1.78	34.59	2.46	42.42	1.09	8.20	1.28	14.11	1.81	12.67	1.43	23.21
n-pentane	0.86	15.21	1.12	18.28	0.51	3.81	0.56	6.17	0.63	4.35	0.53	7.12
1,3-butadiene	0.23	3.47	0.28	5.62	0.12	0.67	0.12	1.72	0.13	1.37	0.10	0.94
trans-2-pentene	0.13	2.71	0.18	3.68	0.08	0.55	0.10	1.39	0.10	2.14	0.10	1.44
cis-2-pentene	0.07	1.41	0.10	1.91	0.05	0.26	0.05	0.66	0.05	0.99	0.05	0.66
2-methylpentane	0.65	14.70	0.93	18.84	0.36	2.31	0.40	4.79	0.57	8.42	0.45	3.82
3-methylpentane	0.31	6.41	0.46	8.53	0.18	1.18	0.20	2.41	0.29	4.07	0.21	1.86
isoprene	0.16	1.93	0.18	2.61	0.08	0.45	0.09	0.95	0.11	0.91	0.08	0.60
n-hexane	0.29	4.93	0.38	6.12	0.15	0.87	0.15	1.82	0.19	2.00	0.14	1.37
n-heptane	0.22	3.75	0.21	2.13	0.09	0.39	0.09	0.82	0.11	0.98	0.08	0.64
benzene	1.23	18.78	1.66	17.53	0.71	4.02	0.76	8.17	0.90	7.85	0.61	5.34
toluene	2.26	44.55	3.00	39.05	1.28	9.25	1.49	17.57	1.87	16.68	1.36	12.40
ethylbenzene	0.68	7.77	0.84	7.31	0.30	1.62	—	—	—	—	0.08	0.60
1,3- and 1,4-dimethylbenzene	1.63	25.53	1.96	23.92	0.97	5.24	1.05	9.63	1.23	8.76	0.94	6.02
1,2-dimethylbenzene	0.55	9.60	0.66	8.70	0.32	1.82	0.36	3.52	0.40	3.28	0.34	2.01

Table 4 *continued*

Compound	July		August		September		October		November		December	
	average	*hourly max.*	*average*	*hourly max.*	*average*	*hourly max.*	*average*	*hourly max.*	*average*	*hourly max.*	*average*	*hourly max.*
ethane	0.76	2.85	0.72	1.94	1.03	5.69	1.59	6.90	1.66	9.72	1.86	16.82
ethene	0.90	4.74	1.04	8.37	1.80	17.56	2.79	37.33	2.69	23.18	3.11	47.61
propane	3.05	52.57	2.79	33.01	4.50	49.03	6.50	411.15	25.67	386.17	25.57	458.77
propene	1.03	3.73	0.62	4.80	1.37	68.43	1.78	22.88	1.77	15.34	1.83	29.62
ethyne	2.20	11.98	2.41	20.05	4.24	39.45	6.57	256.01	5.96	90.68	5.14	147.94
2-methylpropane	1.12	12.28	1.15	10.11	1.79	56.94	2.27	18.59	2.34	18.90	2.38	31.33
n-butane	2.27	24.07	2.37	21.80	4.04	57.82	5.11	82.56	5.07	65.60	4.48	87.85
trans-2-butene	0.25	0.88	0.21	1.08	0.25	2.37	0.27	4.39	0.21	1.86	0.22	4.63
1-butene	0.12	0.71	0.17	1.49	0.28	3.25	0.33	5.72	0.27	2.43	0.34	6.84
cis-2-butene	0.11	0.64	0.11	0.79	0.16	1.74	0.18	3.45	0.14	1.49	0.15	3.32
2-methylbutane	1.49	11.00	1.61	13.01	2.28	24.26	3.10	75.32	2.92	37.53	1.76	49.91
n-pentane	0.59	4.35	0.64	4.97	0.97	10.14	1.24	21.86	0.99	10.75	0.97	18.07
1,3-butadiene	0.21	1.39	0.26	4.31	0.44	4.72	0.61	9.51	0.59	6.44	0.45	11.51
trans-2-pentene	0.10	0.69	0.12	0.89	0.18	2.13	0.23	6.51	0.20	2.32	0.16	3.29
cis-2-pentene	0.05	0.31	0.06	0.56	0.09	1.10	0.12	3.36	0.10	1.20	0.09	1.70
2-methylpentane	0.29	1.98	0.32	2.90	0.66	6.68	0.78	15.83	0.66	7.61	0.63	13.49
3-methylpentane	0.30	1.30	0.35	3.67	0.49	4.86	0.78	10.93	0.59	5.69	0.46	10.30
isoprene	—	—	0.09	0.52	0.13	1.23	0.16	2.00	0.14	1.39	0.14	2.11
n-hexane	0.19	0.88	0.22	1.71	0.35	3.53	0.46	8.09	0.36	3.38	0.38	6.51
n-heptane	0.10	0.45	0.13	0.93	0.20	1.54	0.26	3.15	0.21	3.43	0.22	3.47
benzene	0.61	3.23	0.67	5.72	1.08	10.45	1.53	23.55	1.42	14.08	1.57	26.50
toluene	1.54	8.23	1.73	14.87	2.77	28.05	3.57	61.50	3.01	30.00	3.08	59.90
ethylbenzene	0.36	5.38	0.43	3.65	0.72	4.78	0.92	10.03	0.82	4.82	0.83	9.94
1,3- and 1,4-dimethyl-benzene	0.90	6.67	1.16	8.01	1.84	15.85	2.40	34.15	2.05	16.49	1.92	34.97
1,2-dimethylbenzene	0.59	3.21	0.66	2.66	0.87	6.17	0.93	12.05	0.75	6.40	0.76	11.85

All provisional data. — = No data available.

Table 5 Annual means for all compounds and sites, January–December 1994 (ppbv)

Compound	Birmingham	Cardiff	Edinburgh	Bristol*	London 1**	London 2	Middlesbrough	Belfast
ethane	5.41	6.47	3.75	1.24	1.99	3.41	2.06	1.46
ethene	3.03	3.55	1.87	1.04	2.57	3.13	1.70	1.78
propane	3.18	2.39	2.61	1.84	2.75	1.61	5.02	7.69
propene	1.90	1.63	1.04	1.64	2.46	1.37	2.76	1.29
ethyne	4.76	4.21	2.79	3.62	8.27	2.53	8.20	3.37
2-methylpropane	1.77	2.27	1.64	1.78	3.24	2.42	3.84	1.65
n-butane	3.86	4.21	2.75	2.83	6.36	3.16	7.38	3.36
trans-2-butene	0.38	0.72	0.20	0.29	0.39	0.23	0.80	0.23
1-butene	0.24	0.27	0.15	0.26	0.53	0.21	0.60	0.21
cis-2-butene	0.22	0.37	0.12	0.25	0.30	0.16	0.62	0.14
2-methylbutane	2.44	3.16	1.59	2.43	4.31	2.03	5.02	1.91
n-pentane	0.71	0.96	0.79	0.89	1.13	0.68	2.20	0.80
1,3-butadiene	0.26	0.28	0.19	0.80	0.62	0.27	0.28	0.30
trans-2-pentene	0.21	0.23	0.14	0.29	0.42	0.22	0.76	0.14
cis-2-pentene	0.11	0.15	0.07	0.17	0.22	0.12	0.36	0.07
2-methylpentane	0.75	0.83	0.33	0.75	1.03	0.72	1.27	0.56
3-methylpentane	0.42	0.44	0.26	0.68	0.78	0.40	0.77	0.37
isoprene	0.23	0.38	0.12	0.33	0.33	0.24	0.57	0.13
n-hexane	0.23	0.37	0.31	0.30	0.31	0.21	0.61	0.27
n-heptane	0.25	0.27	0.15	0.24	0.34	0.18	0.41	0.16
benzene	1.12	1.41	0.69	0.96	1.82	1.14	1.35	1.06
toluene	2.54	2.60	1.75	2.15	3.92	2.02	3.23	2.24
ethylbenzene	1.35	0.67	0.76	0.91	1.53	0.63	2.27	0.66
1,3- and 1,4-dimethyl-benzene	1.72	1.16	1.15	1.87	2.10	0.91	1.76	1.52
1,2-dimethylbenzene	0.93	0.54	0.45	0.75	0.83	0.53	0.99	0.61

Provisional data. * 9 months data only (April–December). **Roadside site.

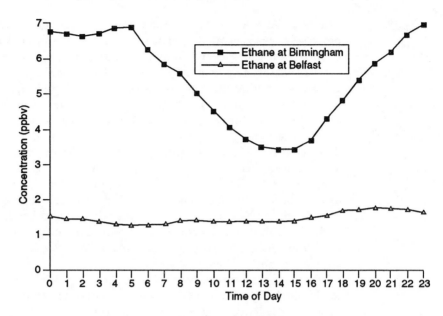

Figure 4 Diurnally averaged ethane concentrations at Birmingham and Belfast for 1994

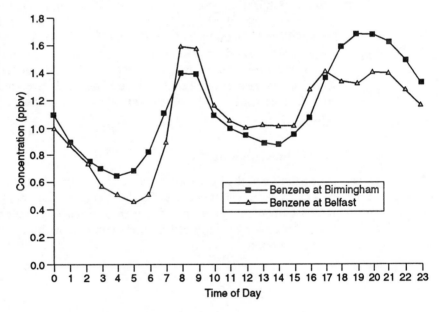

Figure 5 Diurnally averaged benzene concentrations at Birmingham and Belfast for 1994

are taken relatively infrequently, compared to the continuous sampling of the VOCAIR. Additionally, most UK mainland sites are influenced by urban plumes, whereas the prevailing wind direction at Belfast gives rise to air of Atlantic origin that has been subjected to almost no influence by natural gas leakage.

The annual diurnal plot for benzene at Birmingham and Belfast (Figure 5) exhibits a typical pattern for a pollutant emitted from a motor vehicle source. At 6 to 7 o'clock in the morning concentrations climb with the onset of the morning rush hour. At 9 to 10 o'clock traffic flow has peaked and atmospheric mixing processes begin, generated by increased solar radiation. This leads to an effective dilution and dispersion of the pollutant and concentrations decline until the onset of the evening commutor activity when concentrations again increase. During this period atmospheric mixing is suppressed, again contributing to pollutant accumulation. Following the passing of the evening traffic peak flow and onset of nocturnal stability the concentration of benzene slowly declines through the night. This is probably due to a combination of weak dispersion and deposition at the surface. Although both sites show consistent patterns for benzene they differ in terms of the shapes of the morning and evening peaks. It is likely that peak traffic flows in Birmingham build over a longer period than is the case for Belfast. The morning peak for Belfast is well defined and occurs for a shorter period. The evening peak at Birmingham is also significantly higher than at Belfast. This may be a result of the extensive motorway network surrounding the Birmingham site; this network carries a large volume of traffic over an extended period of the day.

More reactive hydrocarbons of motor vehicle origin, *e.g.* 1,3-butadiene, will exhibit similar patterns to benzene but ambient concentrations will be influenced strongly by reaction with ˙OH free radicals in the daytime atmosphere.

Recent publications from the UK Expert Panel on Air Quality Standards have recommended standards for the two carcinogenic compounds benzene and 1,3-butadiene.[2,3] In the case of benzene this is 5 ppb as a rolling annual mean with a target of 1 ppb, and for 1,3-butadiene 1 ppb as a rolling annual mean. The data from Table 5 for benzene indicates that for 1994 the annual means were well below the 5 ppb threshold; Edinburgh and London 2 were below the 1 ppb target value. In all cases the 1,3-butadiene annual means were below the recommended standard.

4 Acknowledgements

The operation of the hydrocarbon network is funded by the UK Department of the Environment under contract PECD 7/12/115. The authors are grateful for the assistance of many site operators in the maintenance of the network sites. The data reported for Exhibition Road were collected with the support of Imperial College, London.

Source Inventories and Control Strategies for VOCs

NEIL R. PASSANT

1 Introduction

The major sources of VOC emissions are organic solvents, the oil and chemical industries, combustion sources, motor vehicles, and natural sources. Other minor sources of VOCs include the metals industries, food and drink manufacture, waste disposal, and straw and stubble burning.

The major use of solvents is in the formulation of industrial coating materials such as paints, inks, and adhesives. Evaporation of the solvent after application is an essential aspect of their function. To provide safe working conditions within workplaces, these solvent emissions are usually contained and vented to atmosphere from stacks, although this is not possible when coating is carried out in the open. During all coating processes, some part of the total emission is not contained and this is often referred to as a 'fugitive' emission; handling and storage of coating materials, washing of equipment, and spillages can all cause this type of emission. The scale of coating operations varies considerably. Solvents are also used in other industrial processes, for instance the extraction of vegetable oil from oil seeds, the manufacture of pharmaceuticals, the cleaning of metal and plastic components, and 'dry cleaning' of textiles. Another major use of solvents is in the formulation of products for use by consumers, for instance aerosols, cosmetics, and household paints. A wide range of organic solvents is used, including aromatic and aliphatic hydrocarbons, oxygenates such as alcohols, ketones, esters, and chlorinated hydrocarbons.

Crude oil production is a major emission source in some countries. Flaring and venting of gases from production facilities, together with displacement of vapour during loading of crude oil into tankers are the major sources. Oil refineries and chemical plants are large sources of VOCs. Emissions occur from storage of volatile materials, venting and flaring of gases from processes, from effluent streams, and fugitive emissions from valves, flanges, and pumps. The distribution of motor spirit is a further major source, emissions occurring during loading of road tankers and filling station storage facilities ('stage I emissions') and during vehicle refuelling ('stage II emissions'). Compounds emitted from oil industry sources are predominantly aliphatic hydrocarbons, while chemical processes will emit all classes of VOCs.

Emissions from vehicles can be divided into emissions due to incomplete fuel combustion (exhaust emissions) and losses of fuel prior to combustion (evaporative emissions). Emissions from stationary combustion sources are much less significant. Combustion processes emit both aliphatic and aromatic hydrocarbons, together with some oxygenated species.

Finally natural biogenic sources, particularly trees, contribute to VOC emissions. Estimates of biogenic emissions vary considerably; nevertheless, natural emissions represent a high proportion of overall emissions in some regions.

Clearly, there is a wider range of sources of VOCs than is the case for most other common air pollutants such as NO_x or CO (emitted from combustion sources) and this makes the compilation of an emissions inventory particularly difficult. Estimates need to be made for a large number of sources and, due to the unique nature of many sources, measurements made for one process cannot easily be used to derive emission estimates for other processes.

The majority of western hemisphere industrial nations now produce inventories of air pollutants, although VOCs are not always included. A recent survey of national inventories[1] found that of 18 national inventories, less than half included VOC emissions. In the UK, emissions of non-methane VOCs (NMVOC), methane, and other atmospheric pollutants are estimated for inclusion in the Department of the Environment's (DoE) National Atmospheric Emissions Inventory which is published annually in the 'Digest of Water Protection and Environmental Statistics' by HMSO.

Since many pollution problems are trans-boundary, inventory development increasingly is being co-ordinated on an international basis. Early work was carried out by the Organisation for Economic Co-operation and Development (OECD) and the Commission of the European Communities (CEC), based largely on emission estimates taken from national inventories. Subsequently, the CEC produced a handbook of default emission factors as part of the CORINAIR 85 project. A later project, CORINAIR 90, is revising the default emission factor handbook, including data quality assessment and developing a methodology for handling chemical speciation of VOC emissions. In order to develop more coherent inventories for an even wider geographical area the development of an emissions inventory guidebook has been undertaken by the United Nations Economic Commission for Europe (UN ECE). This project complements the CORINAIR project and the UNEP greenhouse gas inventory work. A summary of these and other recent international inventory projects has been given elsewhere.[2]

This international inventory work indicates that the relative importance of the various sources differs considerably from country to country, as illustrated in Table 1.

---

[1] G. McInnes, 'Review of National Emissions Reports and Methodologies', in Proceedings of the 2nd Meeting of the Task Force on Emission Inventories, Delft, The Netherlands, 1993; ed. G. McInnes, J.M. Pacyna, and H. Dovland (NILU, Lillestrom, Norway), 1993. *See also* EHEP/CCC Report 8/93.

[2] S.J. Richardson and M. Woodfield, 'International VOC Inventories', in Proceedings of the VOC Assessment and Evaluation Workshop, Amersfoort, The Netherlands, 1993; ed. H. Bloeman and J. Burn, Dutch Institute of Public Health and Environmental Protection (RIVH, Bilthoven, The Netherlands), 1993.

Table 1 Comparison of the relative importance of VOC sources in selected countries

Emission source	% of total NMVOC emissions*		
	Austria[3]	Norway[4]	UK[5]
Solvent use	36	12	28
Oil industry	2	44	16
Chemical industry	<1	<1	7
Stationary combustion	18	4	2
Other stationary sources	10	2	5
Mobile sources	29	38	37
Natural sources	—	—	3

*NMVOC: non-methane volatile organic compounds.

2 Developing Inventories

The basis of most emission inventories is the use of 'emission factors' which relate emissions of a given pollutant to the level of the activity giving rise to that emission. For example, emissions of solvent during the coating of automobiles might be expressed in terms of kg solvent emitted per vehicle coated, or as an emission per unit of paint consumed. In general, therefore, emissions are calculated using the relation:

$$\text{emission} = \text{emission factor} \times \text{activity statistic}$$

Emission factors can be derived from direct measurement of sources; however, given the large diversity of sources, this is costly. Emission factors are, therefore, often based on rather limited data and the results of a single measurement may be applied very widely.

An alternative bottom-up approach to developing an inventory has been used in The Netherlands. This involved the collection of emissions data from a large number of individual installations (20 000); subsequently, a smaller subset (700), which were responsible for the greater part of the emissions, were chosen to be sampled and inventoried frequently. This approach can only be successful with the co-operation of industry, and relies upon plant operators having a good understanding of their emissions. In the UK, emissions data are now becoming available due to the requirement of the Environmental Protection Act 1990 that many industrial plants register their processes in order to obtain authorization to operate. Emissions data for processes controlled under Integrated Pollution Control have already been gathered by Her Majesty's Inspectorate of Pollution to form a Chemical Release Inventory. These sources of information will undoubtedly be of great importance in refining the National Atmospheric Emissions Inventory in the future.

Basic inventories give estimates of total VOC emissions for each recognized

[3] W. Loibl, R. Orthofer, and W. Winiwarter, 'Spatially Disaggregated Emission Inventory for Anthropogenic NMVOC in Austria', *Atmos. Environ.*, 1993, **24A**, 2575.

[4] Central Bureau of Statistics, 'National Resources and the Environment 1992', Oslo, 1993, ISBN 82-537-3855-2.

[5] C. A. Gilham, S. Couling, P. K. Leech, H. S. Eggleston, and J. G. Irwin, 'UK Emissions of Air Pollutants 1970–1991', Warren Spring Laboratory, Report LR 961, 1993.

Table 2 Selected estimates of VOC sources in the UK

Process type	Emission (kilotonnes)	Number of sources
Chemicals manufacture	200	>300
Oil refineries	100	14
Coating of metals and plastics	98	1.2×10^4
Aerosols	86	7.7×10^8
Surface cleaning	43	2.5×10^4
Printworks	41	3.7×10^3
Whisky distilleries	40	114
Coatings – vehicle refinish	22	1.6×10^4
Dry cleaning machines	10	4.5×10^3
Power stations	6	167

source. Increasingly, fuller speciation and details of spatial and temporal disaggregation are being included. Speciation of VOC inventories is of importance due to the differing environmental impacts of individual VOC species. Models of photochemical pollution require speciated emissions data, although simplified schemes are often used. Improved speciation of inventories is needed so that the validity of these simplified schemes can be verified and more detail included in future models. Speciation of the UK VOC inventory has reached a good stage of development. A recent study[6] provided species profiles for 85% of UK emissions and identified over 300 compounds or groups of compounds (such as white spirit) present in emissions. A small number of compounds, mainly alkanes and aromatic hydrocarbons, are responsible for over 50% of the mass emission.

Good progress towards estimating spatial disaggregation of VOC inventories has also been achieved. In principle, emissions data are collected for each source which is then treated in the inventory as arising at a particular geographical location. However, there are a large number of these 'point sources' and, in compiling a spatially disaggregated inventory, a decision has to be made about the number of point sources that can be identified. The remaining emission is disaggregated using surrogate statistics such as population, employment, or traffic statistics. Estimates of numbers of point sources for selected emission sources in the UK are shown in Table 2. Although it would be possible for each refinery to be treated as a point source, aerosols would be disaggregated on the basis of population. In the case of many VOC sources, such as printworks, although the average emission would be relatively small (approximately ten tonnes), some plants could emit hundreds or even thousands of tonnes of VOCs annually. Ideally, all point sources which emit more than a set threshold mass emission should be included in a spatially disaggregated inventory.

A spatially disaggregated inventory for Austria[3] treated many industrial plants as point sources and disaggregated all other sources using surrogate statistics and derived emission densities ranging from 0 tonnes $km^{-2} yr^{-1}$ in unpopulated areas to 700 tonnes $km^{-2} yr^{-1}$ along major roads. Inventories such as this rely heavily on the use of surrogate statistics, and emissions data need to be collected for individual plants in order to improve their accuracy.

Temporal disaggregation of emission inventories has received less attention.

[6] H.J. Rudd, 'Emissions of Volatile Organic Compounds from Stationary Sources in the United Kingdom: Speciation', AEA Technology, Report no. AEA/CS/16419033/REMA-029, 1995.

	A	B	C	D	E
SO_2	95	2	0	0	3
NO_x	68	15	11	1	5
VOC	7	9	11	5	68

Table 3 Distribution of emissions in the 1985 US NAPAP inventory by emission factor data quality rating

Ambient measurements have shown variations in the concentrations of individual hydrocarbon species, both seasonally and diurnally. Summer concentrations of hydrocarbons are significantly lower than winter levels; however, this is explained by shorter atmospheric lifetimes in summer than winter, coupled with generally poorer dispersion of pollutants in winter. Some emission sources, such as evaporation of fuels and chemical feedstocks from tanks, should show an increased emission in summer while emissions from combustion (particularly domestic combustion processes) and vehicle exhausts will increase in cold weather. The remaining sources would not be expected to show a very great seasonal variation and overall VOC emission levels might be expected to remain fairly constant throughout the year. It should be noted that emissions of some VOC species, such as benzene, which is of concern due to its toxicity, and which is emitted mainly from motor vehicle exhausts, may show a stronger seasonal variation. Part of the diurnal variations in concentrations of hydrocarbons detected by ambient measurements will be due to emissions from sources such as vehicle emissions, domestic solvent use, and many industrial processes being concentrated in the daytime. Emissions from other sources, such as gas leakage and continuous industrial processes, will remain fairly constant throughout a 24 hour period.

3 Improving and Verifying Inventories

Existing inventories are subject to significant levels of uncertainty and a number of approaches have been used to indicate or quantify this uncertainty. Data quality ratings have been widely used: each emission estimate is assigned a rating ranging from A through to E, with A indicating the highest quality. There is a degree of subjectivity in applying these ratings; nonetheless they are a guide to the reliability of emission estimates and, in the UK, where the DoE has funded research to improve the accuracy of the national VOC inventory, they have been used to highlight where further data is needed. Analysis of emission estimates in the US National Acid Precipitation Assessment Programme (NAPAP) inventory,[7] shown in Table 3, shows that the data quality of VOC emission estimates is considerably lower than for SO_2 or NO_x. VOC emission estimates published by the UK Department of the Environment[8] are mostly assigned C or D ratings.

Although indicative data qualities are a useful tool for prioritizing resources

[7] J.D. Mobley, 'Uncertainty Assessment of Emission Inventories', in Proceedings of the TNO/EURASAP Workshop on the reliability of VOC emission databases, Delft, The Netherlands, 1993; ed. H.P. Baars, P.J.H. Builtjes, M.P.J. Pulles, and C. Veldt (TNO Institute of Environmental Sciences, Delft, The Netherlands), IMW–TNO Publication P93/040, 1993.

[8] Department of the Environment, 'Reducing Emissions of Volatile Organic Compounds (VOCs) and Levels of Ground Level Ozone: A UK Strategy', 1993.

Figure 1 Uncertainty in
the UK methane inventory

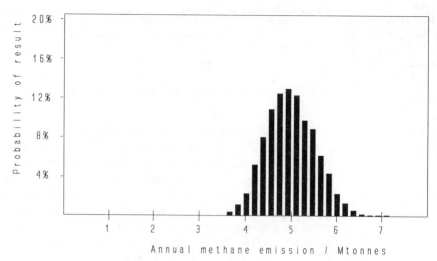

for further data collection, quantification of uncertainties in inventories is needed
by modellers and policy makers. Some inventories have included upper and
lower limits to emission estimates. However, a lack of data has often meant that
these limits cannot be quantified sufficiently accurately to allow a useful
sensitivity analysis. For example, where an emission estimate is based on a single
emission factor, upper and lower limits cannot be derived; however, the
uncertainty would actually be high. There is also a difficulty in determining the
uncertainty in total emissions. Simply summing the lower and upper limits for
each source to obtain an overall lower and upper limit is likely to greatly
overestimate the actual uncertainty.

Distribution functions (such as normal distributions for example) have also
been used to estimate uncertainties in inventories. The inputs used to calculate
emissions from each source (*i.e.* emission factors, activity statistics) can be
entered as probability distributions rather than as single values. Values can then
be sampled from within these distributions and the emissions calculated. The
process is then repeated many times in order to build up a probability function for
the result. This approach has been used to determine uncertainties in the UK
inventory of mobile VOC emissions[9] and the inventory of methane emissions.[10]
An example of the output from this type of analysis is shown in Figure 1. As with
upper and lower limits, however, this approach requires a good data set.

A variety of methods exist by which the accuracy of inventories can be verified
and improved. The simplest approach to verifying emission estimates would
involve generating emission estimates by two or more alternative approaches and
comparing the results. A similar result obtained by both methods would add

[9] H. S. Eggleston, 'Uncertainties in the Estimates of Emissions of VOCs from Motor Cars', in reference 7.
[10] J. R. Bellingham, M. J. T. Milton, P. T. Woods, N. R. Passant, A. J. Poll, S. Couling, I. T.
Marlowe, M. Woodfield, J. Garland, and D. S. Lee, 'The UK Methane Emissions Inventory: A
Scoping Study on the Use of Ambient Measurement to Reduce Uncertainties', National Physical
Laboratory Report DQM 98, 1994.

credibility to the final emission estimate. Comparisons can also be made between emissions data from one country or region with those from another, on a *per capita* basis, for instance, or by comparing emission factors. It should be stressed, however, that there are often real differences in emissions from a particular source between one country and another, due to the use of different technologies or the presence of control standards, for instance, and such comparisons should therefore be treated cautiously. Comparisons between inventories are considerably more useful as checks that all significant sources have been considered.

Monitoring of emission sources can also be used to verify emission estimates and, although the resulting increase in data quality is likely to be higher than that obtained by comparing different estimates, the cost of monitoring can be high. Direct monitoring of emission sources is most suitable for stationary sources such as large combustion plant and industrial plants using solvents. In cases where emission sources are either very numerous or else where emission sources are ill-defined, such as emissions from refinery storage tanks or fugitive emissions from petrochemical plants, indirect monitoring techniques, such as the measurement of pollutant levels in the ambient atmosphere close to the emission sources, can be used. Indirect measurements of emissions from refinery storage tanks have prompted modifications to the estimation methods used by the oil industry. Improvements can also be made to inventories by measurements of process parameters other than emissions; for instance, by quantifying movements of products within a process so as to allow an accurate mass balance to be calculated.

Finally, ambient measurements of hydrocarbons, ozone, and its precursors, may also provide a useful comparison with inventory data. For example, measured ambient levels of hydrocarbons can be compared with theoretical levels calculated by modelling the transport and removal of estimated emissions through reaction with hydroxyl radicals. A study of UK emissions[11] suggests that emissions of hydrocarbons are underestimated by between 5 and 35%. The compounds included in the study are emitted predominantly from vehicles, petrol distribution, and oil refineries. At present, very few data are available on ambient concentrations of other VOC species such as oxygenates and chlorinated species; therefore, no conclusions can be drawn about the accuracy of other parts of the UK emission inventory using this type of analysis.

4   Uses for Inventories

As discussed above, inventories are needed in order to quantify the effect of VOC emissions on the environment. Another major use of inventory data is in the development and monitoring of air pollution control policies to reduce emissions. The data requirements of those engaged in policy development are somewhat different from those of modellers. A knowledge of emissions and an understanding of their contribution to pollution problems is a pre-requisite to the development of a strategy to control the problem effectively; however, additional data are needed to develop forecasting and cost/benefit analysis tools. These data requirements include:

[11] Photochemical Oxidants Review Group (PORG), 'Ozone in the United Kingdom 1993', Department of the Environment, UK, 1993.

- characteristics of stack emissions, *i.e.* volume flowrates and VOC concentrations
- the relative proportions of fugitive and stack emissions from industrial processes
- size distributions of sources
- control techniques which can be applied, and their effectiveness
- the costs associated with control techniques

Forecasts of future trends in VOC emissions have been made since the early 1980s, and have been of great importance in the formulation of national strategies, such as those adopted in The Netherlands and embodied in international agreements such as the UN ECE Protocol (with set targets for emission reductions). Forecasts can be used to check that targets are realistic and the merits of different regulatory approaches can be compared.

Cost/benefit analyses are of great importance, since different strategies may achieve environmental benefits at widely different costs. A good example is a study, commissioned by the Department of the Environment,[12] to examine the costs associated with the proposed EU controls on emissions during vehicle refuelling (the so-called 'Stage II' Directive). The options for control are either to fit vehicles with large carbon canisters (LCCs) to recover vapour, or to recover vapour at the filling station pump and return it to the storage tank. The cost of LCCs (£2723 per tonne abated) was considerably greater than the costs associated with vapour recovery at the filling station (£369 per tonne abated) although greater reductions in emissions could be achieved by fitting LCCs (overall reduction in emissions of 90% compared with 59%). Cost/benefit studies have also recently been carried out to investigate the costs and benefits of the proposed 'Solvents Directive' for each EU member state and for each type of process controlled by the Directive. This study highlights the paucity of data on the numbers and sizes of industrial processes throughout the EU and greater emphasis needs to be placed on the collection of this type of information in the future.

5 Techniques for VOC Abatement

Reductions in VOC emissions from stationary sources can be achieved by four distinct approaches, either singly or in combination:

1. *Resource management* – improving the management and control of processes to minimize emissions and wastage
2. *Product reformulation* – the reduction or elimination of organic solvents from products such as coatings used in a process; options include high solids–low solvent coatings, water-based coatings, radiation-cured coatings, and powder coatings
3. *Process modification* – reducing emissions from a process by improving or otherwise making modifications to the equipment used

---

[12] Chem Systems Ltd, 'Gasoline Vapour Recovery Costs', Report prepared for the Department of the Environment, UK, 1994.

4. *End-of-pipe technologies* – the treatment of emissions from processes by means of specialized technologies which either destroy or capture the emission

An important example of resource management relates to solvent management. Solvents are often perceived as cheap commodities, and little attention has been paid in the past to limiting fugitive emissions. The Commission of the European Community proposed that plant operators should, therefore, estimate their fugitive emissions by developing a 'Solvent Management Plan'. All inputs of solvent to the plant are quantified and all outputs, other than fugitive emissions, are measured directly or estimated. Fugitive emissions can then be estimated from the difference. It was envisaged that, as plant operators became more aware of where solvent emissions were occurring, options for reducing emissions would be identified. Oil refineries and petrochemical plants also can have high fugitive emissions and these can be reduced through the adoption of rigorous maintenance programmes. The benefits of resource management techniques are likely to be highly site specific; however, it is generally thought that such measures can reduce emissions of VOCs, often at low cost and sometimes with cost savings. However, other complementary measures are necessary in order to reduce emissions substantially.

VOCs are frequently used in 'products' as solvent, carriers, or propellants. Thus a relatively well established avenue for control is to seek to remove, reduce, or alter the VOC component in these products. Options range from solvent-free systems through to water-borne and high-solids systems where the weight of solvent has been considerably reduced from conventional systems. It should be noted that water-borne products are seldom VOC free: water-borne coatings can contain up to 25% organic solvents, and they also tend to have lower solids contents.

The environmental benefits of switching from one system to another may be, to a certain extent, replaced by other environmental problems, *i.e.* increased energy use or waste disposal problems. An assessment of all the environmental impacts of new products therefore is desirable. For many products containing solvents, particularly those products for domestic use, reformulation is the only viable option to reduce emissions. Product reformulation may also be more suitable for small scale industrial solvent users, to avoid the high costs associated with end-of-pipe technologies.

Where VOCs are used as a primary raw material, as fuels or feed material for a chemical process, then losses can occur in waste streams and as 'fugitive' losses from storage containers, pipes, and transfer systems, and from reaction or combustion vessels. Process modifications may be considered in order to reduce emissions; these can vary between the use of improved fittings such as valves and pumps to a complete redesign of a process. Similarly, both the timescale over which process changes can be made and the scale of environmental benefits which accrue vary considerably. In certain sectors, such as the petrochemical industry, which have significant fugitive emissions, process changes probably represent the best long-term solution for reducing emissions.

End-of-pipe abatement systems can broadly be classified into two groups – those that destroy the VOC and those that recover the VOC from the waste stream. Incineration is the most widely used destructive technology. Incinerators can be split into thermal systems which raise the temperature of waste gases typically to above 750 °C, and catalytic systems, which employ platinum group metals to reduce the temperature at which destruction occurs. Incineration can be used to treat all VOCs, although chlorinated hydrocarbons can be oxidized to form acidic by-products, necessitating further treatment of waste gases.

Biological degradation makes use of the ability of certain micro-organisms to break organic compounds down. The most commonly used form of biological degradation is the biofilter, which consists of a bed of material such as heather, bark, or peat through which the waste gas flows. The bed is kept moist and a population of micro-organisms develops in a 'biofilm', consisting of water and micro-organisms, on the surfaces of the packing material. Organic pollutants dissolve into the water where the micro-organisms can degrade them. The pollutants have to be at least slightly soluble in water and biodegradable. Compounds which occur naturally, such as alcohols and alkanes generally degrade well, whereas compounds such as chlorinated hydrocarbons are difficult to treat in biofilters. Biofilters can be an inexpensive abatement technology; however, their efficiency may be low.

The most widely used technology for recovery of VOCs is adsorption, usually with activated carbon, although other adsorbents such as zeolites can be used. Molecules of the waste gas adhere to the surface of a solid (adsorbent). Adsorption is essentially a batch process since the capacity of the adsorbent is limited. Regeneration of the adsorbent (*i.e.* desorption of the organic pollutant, which can then be recovered and reused) is usually effected by either heating the adsorbent or by stripping with steam. A variety of adsorption systems are available. In some, multiple beds are used, a proportion undergoing regeneration whilst the remainder are adsorbing. Other systems consist of single units in which the adsorbent is cycled through adsorption and then desorption sections. Adsorption systems can be used to treat most VOCs although they are most effective when treating waste streams containing single compounds or relatively simple mixtures of compounds.

Further recovery techniques include absorption and condensation. Absorption involves the transfer of a soluble gas molecule to a solvent liquid such as water or low volatility hydrocarbons. Absorption systems can treat waste gases containing very high concentrations of VOCs. Condensation also works well at high VOC concentrations. The technique is most applicable for organic pollutants having reasonably high boiling points relative to ambient conditions.

Recovery of VOCs enables them to be recycled. The recycling of solvents may be divided into high efficiency recycling of high purity solvents for re-use in the same process, or lower efficiency recycling of solvents, for use in progressively lower value products. In some cases, the recovered VOC is of such a low value, or is so contaminated by the process, that it has no sale value to the operator, however, there is still the possibility that it can be recovered and used to supplement, or even replace, fuels used for process or space heating.

End-of-pipe technologies can achieve extremely high abatement efficiencies; however, they may be very expensive, particularly for small plants, and they cannot be used to treat fugitive emissions.

Abatement of petrol-engined vehicle exhaust emissions relies upon the fitting of three-way catalysts which oxidize carbon monoxide and VOC emissions and reduce NO_x emissions to carbon dioxide, water, and nitrogen, respectively. Evaporative losses from petrol-engined cars can be controlled through the use of carbon canisters: vapours from the principal sources are vented through the canister and, during engine running, the adsorbed hydrocarbons are purged from the canister and passed through the engine. Petrol losses which occur during refuelling may also be treated by an on-board recovery system consisting of a large carbon canister (LCC) through which vapour displaced from the fuel tank during refuelling is passed. A cheaper control technology for refuelling emissions is vapour recovery at the filling station (see page 58). Evaporative losses can also be reduced by decreasing the volatility of fuels.

Control technologies for diesel-engined vehicle exhaust emissions include oxidation catalysts and modifications to the engine design to improve the combustion process.

6 VOC Control Strategies

Most industrial nations are working towards a national reduction target. The UN ECE Protocol for the control of emissions of VOCs was signed in 1991. Under this Protocol participating countries undertook to reduce emissions by 30% by the year 1999, with 1988 generally as the baseline. Some countries have set more stringent targets; for instance, in Austria a reduction of 70% by 2006, based on 1988 levels; and in The Netherlands a 50% reduction by 2000, based on 1981 levels. A number of countries have gone further by also setting reduction targets for individual sectors of industry. The Dutch KWS 2000 control strategy has provided a model which has subsequently been adopted by other countries such as Denmark and Hungary. The Dutch system specifies general measures to be taken to reduce emissions; however, a large degree of flexibility is built into the system. By setting emission reduction targets for branches of industry as a whole, while allowing operators the choice of how to achieve them, it is possible for measures to be targeted to where they are most cost-effective. However, this approach also requires both a high degree of co-operation between regulators and industry, as well as a good knowledge of emission levels and the effectiveness of available abatement technologies.

Other governments have imposed abatement measures on sectors of industry which are major sources of VOCs; examples include Austria and Norway. Five control measures have been proposed in Norway; these are recovery of crude oil vapour during loading onto marine tankers, emission controls in the petrochemical and oil refining industries, recovery of petrol vapour during distribution, and abatement of industrial solvent emissions. The Austrian measures include regulations on the use of solvents, controls on petrol distribution emissions, and promotion of measures to save energy.

The United Kingdom, Germany, and – to a lesser extent – France have developed regulations which generally set emission limits rather than specify a particular abatement measure. The emission limits are expressed in terms of the concentration of VOCs in waste gases and for the bulk of VOC sources these limits are achievable with existing end-of-pipe technologies. Two regimes of control exist in the UK. Emissions to air, land, and water from the most polluting processes are regulated by Her Majesty's Inspectorate of Pollution through Integrated Pollution Control (IPC) and emissions to air from the remaining processes are regulated by Local Authority Air Pollution Control (LAAPC). Emission limits have been generally set at 50 mg m^{-3} for most LAAPC processes, expressed as total carbon. An alternative to meeting emission limits is available for processes which use solvents in coatings. Compliant coatings have been defined which have lower organic solvent contents than traditional coatings, and process operators can switch to the use of these products rather than fit end-of-pipe technology. Emission limits for IPC processes depend upon the compounds present in the emission. A more stringent limit is set on emissions of compounds which are toxic or which cause stratospheric ozone depletion. Emission limits have also been set in the past on the basis of mass flow limits for individual substances, for instance in Denmark. Most non-chlorinated organic solvents have a limit of 6.25 kg per hour.

The USA has had to adopt a flexible approach due both to its federal system of government and because the extent of the pollution problem varies across the continent. A National Ambient Air Quality Standard (NAAQS) for ozone was established by amendments to the Clean Air Act in 1990. Areas were then classified according to whether they attained the standard, and those not complying with the standard were classified according to the severity of the pollution problem. The stringency of emission control is then linked to this classification. For instance, all non-attainment areas must place emission controls on all 'major' stationary sources of VOCs. These major sources vary from sources emitting more than 100 tonnes of VOC per year to those emitting more than 10 tonnes per year in the case of the most extreme non-attainment.

As well as prescribing abatement measures or emission standards, a number of economic instruments which encourage emission reductions by industry are available. A selection of some of the most commonly proposed measures is given below:

- effluent charges paid for emissions into the environment, based on the amount and/or impact of the emitted pollutants
- tax differentiation giving more favourable prices for 'environmentally-friendly' products
- grants to polluters to reduce their pollution levels
- emissions trading allowing a company to sell or trade the difference between its actual emissions and its allowable emissions to another company which then has the right to release more than its normal limit allows
- liability insurance – due to the legal liability of polluters for environmental damage or clean-up costs associated with emissions, insurance premiums will reflect the probable damage/clean-up costs and the likelihood that damage will occur

Reductions in emissions from domestic products can be more difficult to achieve because product reformulation is probably the most effective option. The potential for switching to low-solvent alternatives varies; significant reductions in solvent emissions from decorative paints have been made in the past due to the change from solvent-based to water-based paints. Less scope for reductions exist for cosmetics, aerosols, and other household products and, where alternatives exist, consumers may be reluctant to change to new products. Control of emissions from domestic sources usually is effected, therefore, by a combination of direct controls on a limited number of products and measures to encourage the public to reduce emissions. Direct measures have included controls on the content of products and reduction targets agreed with industry.

A variety of measures to use public pressure as a means of reducing the mass of VOCs emitted from domestic sources have also been tested including the following areas:

- labelling and life-cycle analysis
- air quality bulletins issued to the media
- leaflets/information hand-outs – in the UK these have included a leaflet published by the Department of the Environment entitled 'Summertime Smog', which gives advice to the public on how they can help reduce air pollution, especially during periods of peak ozone formation

A uniform approach to control of motor vehicle emissions has been adopted throughout much of Europe with the implementation of a number of EU Directives setting progressively tighter emission limits for exhaust and evaporative emissions from cars and lorries. The current emission limits can be met by fitting three-way catalytic converters to petrol engined vehicles together with small carbon canisters, and emission limits for diesel engine vehicles can be met by improvements in engine design. The emission limits will be progressively tightened as control technology develops.

The reductions in emissions that will result from these measures will be partially lost in the longer term due to the anticipated increases in traffic and congestion. Thus, Governments are also looking at measures to control growth in traffic. The eighteenth report of the Royal Commission on Environmental Pollution made a large number of recommendations on the measures that should be taken to reduce pollution associated with transport. Major themes which were covered include the following areas:

- continued tightening of vehicle emission limits to force improvements in emission controls
- fiscal incentives to encourage more rapid introduction of pollution control devices to new vehicles and retrofitting to existing vehicles
- more stringent emissions testing for vehicles
- increased duty on fuel to encourage fuel efficiency
- measures to encourage the use of alternative fuels and to change the composition of existing fuels to reduce emissions
- support for the development of more environmentally friendly means of propulsion
- encouraging more use of public transport and cycling

- reducing expenditure on (a) new road construction and (b) improving the capacity of existing roads

7 Acknowledgements

This article has been written as part of the air pollution research programme 'Measurement and Control of VOCs from Stationary Sources' (contract number PECD 7/12/139) of the UK Department of the Environment. However, all views expressed are those of the author and do not necessarily reflect DoE policy.

Gas Phase Tropospheric Chemistry of Organic Compounds

ROGER ATKINSON

1 Introduction

The troposphere is that region of the Earth's atmosphere into which chemical compounds are generally emitted as a result of human activities (an exception being the exhaust from present and future supersonic transports). As described in the Sections below, the troposphere is a chemical 'processor' resulting in the partial or complete degradation of almost all organic compounds emitted into the atmosphere, with the exception of the chlorofluorocarbons (CFCs) and several of the bromochlorofluorocarbons (Halons). The CFCs and Halons are therefore transported into the stratosphere and are photolysed by short-wavelength ultraviolet radiation as they rise through the stratospheric ozone layer, to produce Cl$^{\bullet}$ or Br$^{\bullet}$ atoms which lead to depletion of stratospheric ozone through the ClO$_x^{\bullet}$ and BrO$_x^{\bullet}$ catalytic cycles.[1,2]

The focus of this chapter is on the gas-phase chemistry of volatile organic compounds in the troposphere. Organic compounds are expected to exist, at least partially, in the gas phase provided that they have liquid-phase (or sub-cooled liquid) vapour pressures of $> 10^{-4}$ Pa at the ambient atmospheric temperature,[3] and this demarkation between gas-phase and particle-associated organic compounds based on their liquid-phase vapour pressure can be used as a first approximation to assess the gas/particle distribution of chemicals in the atmosphere. No discussion is given here of the physical or chemical loss processes of particle-associated chemicals, gas-to-particle conversion, or dry and wet deposition of gas-phase organic compounds. Because of space limitations, the tropospheric chemistry of several important classes of organic compounds, including the hydrochlorofluorocarbons (HCFCs) and hydrofluorocarbons (HFCs) [potential replacements for the CFCs], organosulfur compounds, and polycyclic aromatic hydrocarbons (PAH), is not dealt with here. The tropospheric chemistry of these compound classes has been discussed elsewhere,[1,4-6] and

[1] 'Scientific Assessment of Ozone Depletion: 1991', World Meteorological Organization Global Ozone Research and Monitoring Project—Report no. 25, Geneva, Switzerland, 1992.

[2] J. P. D. Abbatt and M. J. Molina, *Annu. Rev. Energy Environ.*, 1993, **18**, 1.

[3] T. F. Bidleman, *Environ. Sci. Technol.*, 1988, **22**, 361.

[4] G. S. Tyndall and A. R. Ravishankara, *Int. J. Chem. Kinet.*, 1991, **23**, 483.

[5] R. Atkinson, D. L. Baulch, R. A. Cox, R. F. Hampson, Jr, J. A. Kerr, and J. Troe, *J. Phys. Chem. Ref. Data*, 1992, **21**, 1125.

[6] R. Atkinson and J. Arey, *Environ. Health Perspect.*, 1994, **102 (Supplement 4)**, 117.

R. Atkinson

those references should be consulted for further details. The atmospheric chemistry of HCFCs and HFCs is also covered in Chapter 6 (page 91).

Physical and Chemical Characteristics of the Troposphere

The troposphere extends from the Earth's surface to the tropopause at 10–18 km, with the height of the tropopause depending on latitude and season, being highest at the tropics and lowest at the polar regions during wintertime. The troposphere is characterized by generally decreasing temperature with increasing altitude, from an average of 289 K at ground level to 210–215 K at the tropopause. In the atmosphere, pressure decreases monotonically with increasing altitude, from an average of 1013 millibar (mb) at the Earth's surface to 140 mb at 14 km (the average altitude of the tropopause). The lowest kilometre or so of the troposphere contains the planetary boundary layer and inversion layers, with vertical mixing between the boundary and inversion layers and the free troposphere above them being hindered. The troposphere is well-mixed and its composition is 78% N_2, 21% O_2, 1% Ar, 0.036% CO_2, varying amounts of water vapour (depending on altitude and temperature), and minute amounts of a number of trace gases.

Molecular oxygen, O_2, and ozone, O_3, in the stratosphere (the stratospheric ozone layer being centred at an altitude of ~ 25–35 km[1]) absorb ultraviolet radiation below $\lesssim 290$ nm, and hence only radiation from the Sun of wavelength $\gtrsim 290$ nm is transmitted through the stratospheric ozone layer into the troposphere, and impacts the Earth's surface.[7] Any depletion of stratospheric ozone allows shorter wavelength radiation to be transmitted through the stratosphere into the troposphere,[8,9] with potentially adverse effects on the flora and fauna of the Earth's biosphere. Ozone mixing ratios in the stratosphere have maximum values of $\sim 10 \times 10^{-6}$,[1] and there is a net downward transport of ozone from the stratosphere into the troposphere, with destruction of O_3 at the Earth's surface.[10] In addition, there is *in situ* photochemical formation and destruction of ozone in the troposphere.[10,11] The result of downward transport of stratospheric ozone, and *in situ* formation and destruction in the troposphere, is the presence of ozone in the 'clean' natural troposphere.[10] Ozone mixing ratios at 'clean' remote sites at ground level are in the range 10–40 $\times 10^{-9}$ [10,12] and increase with increasing altitude.[10]

Sources of Volatile Organic Compounds

Large quantities of organic compounds are emitted into the troposphere from anthropogenic and biogenic sources.[1,13] Methane, the simplest hydrocarbon, is

[7] G. Seckmeyer and R. L. McKenzie, *Nature (London)*, 1992, **359**, 135.

[8] J. B. Kerr and C. T. McElroy, *Science*, 1993, **262**, 1032.

[9] S. Madronich and F. R. de Gruijl, *Photochem. Photobiol.*, 1994, **59**, 541.

[10] J. A. Logan, *J. Geophys. Res.*, 1985, **90**, 10463.

[11] G. P. Ayers, S. A. Penkett, R. W. Gillett, B. Bandy, I. E. Galbally, C. P. Meyer, C. M. Elsworth, S. T. Bentley, and B. W. Forgan, *Nature (London)*, 1992, **360**, 446.

[12] S. J. Oltmans and H. Levy II, *Atmos. Environ.*, 1994, **28**, 9.

[13] H. B. Singh and P. R. Zimmerman, in 'Gaseous Pollutants: Characterization and Cycling', ed. J. O. Nriagu, Wiley, New York, 1992, p. 177.

emitted into the atmosphere from both biogenic (from wetlands, naturally-occurring natural-gas vents, biomass fires, and termites) and anthropogenic sources (from ruminants, rice paddies, coal and gas exploration, natural-gas pipeline leakages, landfills, and biomass burning).[1,14] The estimated worldwide emissions of methane are ~ 150 million tonnes yr^{-1} from biogenic sources and ~ 350 million tonnes yr^{-1} from anthropogenic sources.[1,14] Large quantities of non-methane organic compounds (NMOC), including isoprene (2-methyl-1,3-butadiene), a series of $C_{10}H_{16}$ monoterpenes and $C_{15}H_{24}$ sesquiterpenes, methanol, *cis*-3-hexen-1-ol, and *cis*-3-hexenyl acetate, are emitted from vegetation.[13,15–18] NMOC are also emitted into the troposphere from a variety of anthropogenic sources, including combustion sources (vehicle and fossil-fueled power plant emissions), fuel storage and transport, solvent usage, emissions from industrial operations, landfills, and hazardous waste facilities. Literature estimates of the US and worldwide emissions of NMOC are ~ 20 million tons yr^{-1} and ~ 100 million tons yr^{-1}, respectively, from anthropogenic sources; and ~ 29 million tons yr^{-1} and ~ 1000 million tons yr^{-1}, respectively, from biogenic sources.[1,17,19] These emissions estimates for NMOC have high uncertainties associated with them (for example, up to a factor of ~ 3 uncertainty for biogenic NMOC emissions in the USA[17]).

In many urban areas, the NMOC composition of ambient air samples is similar to that of gasoline. For example, gasoline in the USA is comprised of ~ 45–60% alkanes, $\sim 10\%$ alkenes, and ~ 30–45% aromatic hydrocarbons,[20,21] while early morning ambient air samples in the Los Angeles, California, air basin are comprised of ~ 45–50% alkanes, ~ 10–15% alkenes, ~ 20–25% aromatic hydrocarbons, and ~ 3–7% carbonyls, with 14% unidentified.[22]

Sources of Oxides of Nitrogen

In addition to emissions of methane and NMOC into the troposphere, oxides of nitrogen (NO_x; $NO_x = NO + NO_2$) are also emitted into, or generated in, the troposphere. NO is emitted from soils and natural fires and is formed *in situ* in the troposphere from lightning,[1,19] and is emitted from combustion processes such as vehicle emissions and fossil-fueled power plants.[19] The estimated US and worldwide emissions of NO_x (including formation from lightning) are ~ 1 million tons yr^{-1} and ~ 10 million tons yr^{-1} (as N), respectively, from biogenic or

14 I. Fung, J. John, J. Lerner, E. Matthews, M. Prather, L. P. Steele, and P. J. Fraser, *J. Geophys. Res.*, 1991, **96**, 13 033.

15 V. A. Isidorov, I. G. Zenkevich, and B. V. Ioffe, *Atmos. Environ.*, 1985, **19**, 1.

16 A. M. Winer, J. Arey, R. Atkinson, S. M. Aschmann, W. D. Long, C. L. Morrison, and D. M. Olszyk, *Atmos. Environ.*, 1992, **26A**, 2647.

17 B. Lamb, D. Gay, H. Westberg, and T. Pierce, *Atmos. Environ.*, 1993, **27A**, 1673.

18 R. C. McDonald and R. Fall, *Atmos. Environ.*, 1993, **27A**, 1709.

19 National Research Council, 'Rethinking the Ozone Problem in Urban and Regional Air Pollution', National Academy Press, Washington, 1991.

20 J. E. Sigsby, Jr, S. Tejada, W. Ray, J. M. Lang, and J. W. Duncan, *Environ. Sci. Technol.*, 1987, **21**, 466.

21 S. K. Hoekman, *Environ. Sci. Technol.*, 1992, **26**, 1206.

22 F. W. Lurmann and H. H. Main, 'Analysis of the Ambient VOC Data Collected in the Southern California Air Quality Study', Report to California Air Resources Board Contract no. A832-130, Sacramento, California, 1992.

natural sources; and ~ 6 million tons yr^{-1} and ~ 40 million tons yr^{-1} (as N), respectively, from anthropogenic sources.[1,19] In urban areas, NMOC and NO_x from anthropogenic sources dominate over NMOC and NO_x from biogenic sources, and the reverse is generally the case in rural and remote areas where, for example, isoprene and, to a lesser extent, monoterpenes dominate over anthropogenic NMOC.

2 Chemical Loss Processes for Volatile Organic Compounds in the Troposphere

The major removal processes for volatile organic compounds in the troposphere are the physical loss processes of dry deposition and wet deposition,[3] which are not discussed here, and the chemical transformation processes of photolysis, reaction with hydroxyl ($^{\cdot}OH$) radicals, reaction with nitrate ($NO_3^{\cdot}$) radicals, and reaction with O_3.[23,24]

Formation of Hydroxyl Radicals

The presence of relatively low levels of O_3 in the troposphere is extremely important, because photolysis of O_3 at wavelengths $\geqslant 290\,nm$ occurs in the troposphere to form the excited oxygen, $O(^1D)$ atom.[5] $O(^1D)$ atoms are then deactivated to ground-state oxygen, $O(^3P)$ atoms, or react with water vapour to generate $^{\cdot}OH$ radicals:[5]

$$O_3 + h\nu \rightarrow O_2 + O(^1D) \qquad (\lambda < 320\,nm) \tag{1}$$

$$O(^1D) + M \rightarrow O(^3P) + M \qquad (M = N_2, O_2) \tag{2}$$

$$O(^3P) + O_2 + M \rightarrow O_3 + M \qquad (M = air) \tag{3}$$

$$O(^1D) + H_2O \rightarrow 2\,^{\cdot}OH \tag{4}$$

At 298 K and atmospheric pressure with 50% relative humidity, $0.2\,^{\cdot}OH$ radicals are produced per $O(^1D)$ atom formed. As shown later in this chapter, the hydroxyl ($^{\cdot}OH$) radical is the key reactive species in the troposphere, reacting with all organic compounds apart from the CFCs and certain of the Halons and acting as a 'garbage disposal' system, or low temperature combustion system.

Based on direct spectroscopic measurements of $^{\cdot}OH$ radical concentrations at close to ground level,[25-34] peak daytime $^{\cdot}OH$ radical concentrations are of the

[23] R. Atkinson, in 'Air Pollution, the Automobile, and Public Health', ed. A. Y. Watson, R. R. Bates, and D. Kennedy, National Academy Press, Washington, 1988, p. 99.

[24] R. Atkinson, *J. Phys. Chem. Ref. Data*, 1994, **Monograph 2**, 1.

[25] F. L. Eisele and D. J. Tanner, *J. Geophys. Res.*, 1991, **96**, 9295.

[26] A. Hofzumahaus, H.-P. Dorn, J. Callies, U. Platt, and D. H. Ehhalt, *Atmos. Environ.*, 1991, **25A**, 2017.

[27] G. H. Mount, *J. Geophys. Res.*, 1992, **97**, 2427.

[28] G. H. Mount and F. L. Eisele, *Science*, 1992, **256**, 1187.

[29] T. M. Hard, A. A. Mehrabzadeh, C. Y. Chan, and R. J. O'Brien, *J. Geophys. Res.*, 1992, **97**, 9795.

[30] C. C. Felton, J. C. Sheppard, and M. J. Campbell, *Atmos. Environ.*, 1992, **26A**, 2105.

[31] F. L. Eisele and J. D. Bradshaw, *Anal. Chem.*, 1993, **65**, 927A.

[32] W. Armerding, M. Spiekermann, and F. J. Comes, *J. Geophys. Res.*, 1994, **99**, 1225.

1×10^6

order of 10^6–10^7 molecule cm^{-3}. A diurnally and annually averaged tropospheric ˙OH radical concentration has also been estimated by comparison of the emissions of methylchloroform (1,1,1-trichloroethane) with its atmospheric concentrations (the major atmospheric loss process of methylchloroform is by reaction with the ˙OH radical),[35] resulting in a diurnally and annually averaged ˙OH radical concentration of 8×10^5 molecule cm^{-3} (24 hr average).[35] Note that ˙OH radicals are formed only during daylight hours.

Formation of Nitrate Radicals

The emission of NO into the troposphere from natural and anthropogenic sources is followed by a series of chemical reactions.[5,19,23] The three reactions

$$NO + O_3 \rightarrow NO_2 + O_2 \tag{5}$$

$$NO_2 + hv \rightarrow NO + O(^3P) \tag{6}$$

$$O(^3P) + O_2 + M \rightarrow O_3 + M \tag{3}$$

interconvert NO, NO$_2$, and O$_3$. Further reaction of NO$_2$ with O$_3$ leads to the formation of the nitrate (NO$_3$˙) radical:

$$NO_2 + O_3 \rightarrow NO_3˙ + O_2 \tag{7}$$

The nitrate radical photolyses rapidly [to form NO + O$_2$ and NO$_2$ + O(^{3}P)], with a lifetime due to photolysis of ~ 5 s for an overhead sun, and reacts rapidly with NO.[5] Hence NO$_3$˙ radical concentrations remain low during daylight hours but can increase to measurable levels during night-time. Measurements made over the past 15 years show night-time NO$_3$˙ radical concentrations at around ground level over continental areas ranging from $<5 \times 10^7$ molecule cm^{-3} to 1×10^{10} molecule cm^{-3} [mixing ratios of $<2 \times 10^{-12}$ to 430×10^{-12}].[36,37]

Photolysis

For a chemical to undergo photolysis in the troposphere, it must absorb radiation in the wavelength range ~ 290–800 nm, with the lower limit being set by transmission of ultraviolet radiation through the stratospheric ozone layer and the upper limit being set by the strength of weak chemical bonds (~ 150 kJ mol^{-1}). Furthermore, having absorbed radiation, the chemical must undergo chemical

[33] D. Poppe, J. Zimmermann, R. Bauer, T. Brauers, D. Brüning, J. Callies, H.-P. Dorn, A. Hofzumahaus, F.-J. Johnen, A. Khedim, H. Koch, R. Koppmann, H. London, K.-P. Müller, R. Neuroth, C. Plass-Dülmer, U. Platt, F. Rohrer, E.-P. Röth, J. Rudolph, U. Schmidt, M. Wallasch, and D. H. Ehhalt, *J. Geophys. Res.*, 1994, **99**, 16 633.

[34] F. L. Eisele, G. H. Mount, F. C. Fehsenfeld, J. Harder, E. Marovich, D. D. Parrish, J. Roberts, M. Trainer, and D. Tanner, *J. Geophys. Res.*, 1994, **99**, 18 605.

[35] R. Prinn, D. Cunnold, P. Simmonds, F. Alyea, R. Boldi, A. Crawford, P. Fraser, D. Gutzler, D. Hartley, R. Rosen, and R. Rasmussen, *J. Geophys. Res.*, 1992, **97**, 2445.

[36] R. Atkinson, A. M. Winer, and J. N. Pitts, Jr, *Atmos. Environ.*, 1986, **20**, 331.

[37] D. Mihelcic, D. Klemp, P. Müsgen, H. W. Pätz, and A. Volz-Thomas, *J. Atmos. Chem.*, 1993, **16**, 313.

nge (dissociation or isomerization). The photolysis rate coefficient, k_{phot}, of an anic compound is given by

$$k_{phot} = \int J_\lambda \sigma_\lambda \phi_\lambda d\lambda \tag{8}$$

ere J_λ is the radiation flux at wavelength λ, σ_λ is the absorption cross-section at velength λ, and ϕ_λ is the photolysis quantum yield at wavelength λ.

Lifetimes of Volatile Organic Compounds in the Troposphere

The potential loss or transformation processes for volatile organic compounds in the troposphere are wet and dry deposition, photolysis, reaction with ˙OH radicals, reaction with NO_3˙ radicals, and reaction with O_3.[23] In addition, there is evidence for the reactions of certain alkanes and alkyl nitrates with Cl˙ atoms in the Arctic troposphere during springtime.[38,39] The overall lifetime, $\tau_{overall}$, of an organic chemical is given by

$$1/\tau_{overall} = 1/\tau_{wet\ dep.} + 1/\tau_{dry\ dep.} + 1/\tau_{phot} + 1/\tau_{OH} + 1/\tau_{NO_3} + 1/\tau_{O_3} + 1/\tau_{Cl} \tag{9}$$

where the lifetime due to photolysis is given by $\tau_{phot} = (k_{phot})^{-1}$ and the lifetimes due to chemical reactions with ˙OH radicals, NO_3˙ radicals, O_3, and Cl˙ atoms are given by, for example, $\tau_{OH} = (k_{OH}[˙OH])^{-1}$, where k_{OH} is the rate constant for the reaction of the chemical with the ˙OH radical and [˙OH] is the ambient tropospheric ˙OH radical concentration.

Rate constants have been measured for a large number of organic compounds with ˙OH radicals,[24,40] NO_3˙ radicals,[24,41] and O_3.[24,42] Rate constants for the reactions of the Cl˙ atom have been measured for a smaller number of organic compounds, and then mainly for alkanes, alkenes, aromatic hydrocarbons, and oxygenated compounds.[5,43–46] The measured rate constants for the ˙OH radical, NO_3˙ radical, and O_3 reactions can be combined with the measured or estimated ambient tropospheric concentrations of ˙OH radicals, NO_3˙ radicals, and O_3 to provide tropospheric lifetimes with respect to the various loss processes. Such calculated lifetimes are given in Table 1 for a series of representative alkanes, alkenes, aromatic compounds, and oxygen- and nitrogen-containing organic compounds.

Further details of the tropospheric reactions of the alkanes, alkenes, aromatic

[38] B. T. Jobson, H. Niki, Y. Yokouchi, J. Bottenheim, F. Hopper, and R. Leaitch, *J. Geophys. Res.*, 1994, **99**, 25 355.

[39] K. Muthuramu, P. B. Shepson, J. W. Bottenheim, B. T. Jobson, H. Niki, and K. G. Anlauf, *J. Geophys. Res.*, 1994, **99**, 25 369.

[40] R. Atkinson, *J. Phys. Chem. Ref. Data*, 1989, **Monograph 1**, 1.

[41] R. Atkinson, *J. Phys. Chem. Ref. Data*, 1991, **20**, 459.

[42] R. Atkinson and W. P. L. Carter, *Chem. Rev.*, 1984, **84**, 437.

[43] R. Atkinson and S. M. Aschmann, *Int. J. Chem. Kinet.*, 1985, **17**, 33.

[44] T. J. Wallington, L. M. Skewes, W. O. Siegl, C.-H. Wu, and S. M. Japar, *Int. J. Chem. Kinet.*, 1988, **20**, 867.

[45] T. J. Wallington, L. M. Skewes, and W. O. Siegl, *J. Photochem. Photobiol., A: Chem.*, 1988, **45**, 167.

[46] S. M. Aschmann and R. Atkinson, *Int. J. Chem. Kinet.*, 1995, **27**, 613.

Table 1 Calculated lifetimes for selected volatile organic compounds with respect to photolysis, reaction with the ˙OH radical, reaction with the NO_3˙ radical, and reaction with O_3[a]

Compound	Lifetime due to			
	˙OH[b]	NO_3˙ [c]	O_3[d]	Photolysis[e]
Propane	13 day		>4500 yr	
n-Octane	1.7 day	250 day	>4500 yr	
Ethene	1.7 day	225 day	10 day	
Propene	7 hr	4.9 day	1.6 day	
Isoprene	2 hr	50 min	1.3 day	
α-Pinene	3 hr	5 min	5 hr	
Benzene	12 day	>4 yr	>4.5 yr	
Toluene	2.4 day	1.9 yr	>4.5 yr	
m-Xylene	7 hr	200 day	>4.5 yr	
Styrene	3 hr	4 hr	20 hr	
o-Cresol	4 hr	3 min	60 day	
Formaldehyde	1.5 day	80 day	>4.5 yr	4 hr
Acetaldehyde	11 hr	17 day	>4.5 yr	6 day
Benzaldehyde	1.1 day	18 day		
Acetone	65 day	>1 yr		∼60 day
2-Butanone	13 day			
Methylglyoxal	10 hr		>4.5 yr	2 hr
Methanol[f]	15 day	≥220 day		
Ethanol	4.4 day	≥50 day		
Methyl t-butyl ether (MTBE)	4.9 day			
Methacrolein	5 hr	∼10 day[g]	15 day	
Methyl vinyl ketone	9 hr		3.6 day	
Methyl hydroperoxide[f]	2.6 day			∼5 day
2-Butyl nitrate	16 day			15–30 day

[a]Rate constants taken from references 24 and 40–42. Where no lifetime is given, this is because of a lack of kinetic data; however, none of the loss processes for which data are not available is expected to be significant.
[b]For a 12 hr daytime average ˙OH radical concentration of 1.6×10^6 molecule cm^{-3}.[35]
[c]For a 12 hr night-time average NO_3˙ radical concentration of 5×10^8 molecule cm^{-3}.[41]
[d]For a 24 hr average O_3 concentration of 7×10^{11} molecule cm^{-3}.[10]
[e]For overhead sun.
[f]Wet and dry deposition also expected to be important.
[g]Assuming a similar rate constant as for crotonaldehyde.

hydrocarbons, and oxygen-containing and nitrogen-containing organic compounds are given Sections 3–7 below. The tropospheric chemistry of these classes of volatile organic compounds has recently been critically reviewed,[24] and that review and evaluation article should be consulted for further details of the kinetics and mechanisms of the tropospheric reactions of these organic compounds.

3 Tropospheric Chemistry of Alkanes

Alkanes do not photolyse in the troposphere, nor react at measurable rates with O_3.[24,42] The alkanes react with ˙OH radicals, NO_3˙ radicals, and Cl˙ atoms, with the ˙OH radical reaction being calculated to generally dominate as the tropospheric loss process (Table 1). As for other saturated organic compounds, these atom and radical reactions proceed by H-atom abstraction from the C–H

71

bonds. For example:

$$\cdot OH + RH \rightarrow H_2O + R\cdot \tag{10}$$

$$NO_3\cdot + RH \rightarrow HONO_2 + R\cdot \tag{11}$$

where RH is an alkane and R⋅ is an alkyl radical. Under all tropospheric conditions, alkyl radicals (including the methyl radical formed from methane) react rapidly and solely with O_2 to form an alkyl peroxyl ($RO_2\cdot$) radical:[5,24]

$$R\cdot + O_2 \xrightarrow{M} RO_2\cdot \tag{12}$$

Alkyl peroxyl radicals can react with NO, NO_2, $HO_2\cdot$ radicals, and $RO_2\cdot$ radicals in the troposphere, with the dominant reaction(s) depending on the relative concentrations of NO, NO_2, $HO_2\cdot$ radicals, and $RO_2\cdot$ radicals. While the self-reactions of alkyl peroxyl radicals and the reactions of alkyl peroxyl radicals with other $RO_2\cdot$ radicals often occur in laboratory systems, the reactions of $RO_2\cdot$ radicals with NO, NO_2, and $HO_2\cdot$ radicals are expected to dominate in the troposphere.[1,10] Reaction with NO leads to the formation of an alkoxyl ($RO\cdot$) radical plus NO_2 and, for $RO_2\cdot$ radicals with ≥ 3 carbon atoms, an alkyl nitrate ($RONO_2$):

$$RO_2\cdot + NO \xrightarrow{M} RONO_2 \tag{13a}$$

$$\longrightarrow RO\cdot + NO_2 \tag{13b}$$

The formation yields of alkyl nitrates from reaction 13a increase with increasing pressure and decreasing temperature, and for secondary $RO_2\cdot$ radicals the alkyl nitrate yields also increase with increasing carbon number in the $RO_2\cdot$ radical.[24,47]

The reactions of $RO_2\cdot$ radicals with NO_2 form alkyl peroxynitrates:[5,24]

$$RO_2\cdot + NO_2 \xrightarrow{M} ROONO_2 \tag{14}$$

However, since the alkyl peroxynitrates rapidly thermally decompose back to NO_2 plus the alkyl peroxyl radical, with thermal decomposition lifetimes of ~ 0.1–1 s at 298 K,[5,24] these $RO_2\cdot + NO_2$ reactions can be neglected in the lower troposphere.

The reactions of $RO_2\cdot$ radicals with $HO_2\cdot$ radicals form hydroperoxides[5,24]

$$RO_2\cdot + HO_2\cdot \rightarrow ROOH + O_2 \tag{15}$$

with the hydroperoxides undergoing wet and dry deposition, photolysis, and reaction with the $\cdot OH$ radical.[1,24] For methyl hydroperoxide (CH_3OOH),

[47] W. P. L. Carter and R. Atkinson, *J. Atmos. Chem.*, 1989, **8**, 165.

photolysis and reaction with the $\cdot$OH radical proceed by[5]

$$CH_3OOH + h\nu \rightarrow CH_3O\cdot + \cdot OH \qquad (16)$$

and

$$\cdot OH + CH_3OOH \begin{cases} \rightarrow H_2O + CH_3OO\cdot & (17a) \\ \rightarrow H_2O + \cdot CH_2OOH & (17b) \\ \qquad\qquad\downarrow \\ \qquad HCHO + \cdot OH \end{cases}$$

respectively, to either form the methoxyl radical ($CH_3O\cdot$) or reform the methyl peroxyl radical ($CH_3OO\cdot$).

The combination reactions of $RO_2\cdot$ radicals proceed by two pathways: one forming the corresponding alkoxyl radical(s) and the other forming an alcohol plus a carbonyl. For example, the self-reaction of the 2-propyl peroxyl radical is shown below:

$$2\,(CH_3)_2CHOO\cdot \begin{cases} \rightarrow 2\,(CH_3)_2CHO\cdot + O_2 & (18a) \\ \rightarrow CH_3C(O)CH_3 + CH_3CH(OH)CH_3 + O_2 & (18b) \end{cases}$$

It is estimated that reaction with NO dominates in the atmospheric boundary layer over continental areas, and in the upper troposphere (because of the higher NO mixing ratios in these airmasses).[1,10] The concentrations of NO are sufficiently low in the lower troposphere over marine areas that reaction with the $HO_2\cdot$ radical dominates.[1,11]

Alkoxyl radicals are therefore formed directly from the reaction of $RO_2\cdot$ radicals with NO, and indirectly (and with generally less than unit yield) through hydroperoxide formation via the $HO_2\cdot$ radical reaction. In the troposphere, alkoxyl radicals can react with O_2 (*e.g.* equations 19 and 20), unimolecularly decompose into two shorter-chain species, or isomerize by a 1,5-H shift through a six-membered transition state.[24] These three reactions are shown in Scheme 1 for the 2-pentoxyl radical.

The alkyl and δ-hydroxyalkyl radicals react further by reaction schemes analogous to those shown above, with the reactions of the $CH_3CH(OH)CH_2CH_2\dot{C}H_2$ radical (formed by isomerization of the 2-pentoxyl radical) forming, in the presence of NO, the δ-hydroxycarbonyl compound 1-hydroxypentan-4-one (Scheme 2). The reactions of α-hydroxyalkyl radicals such as $CH_3\dot{C}(OH)CH_2CH_2CH_2OH$ with O_2 are discussed in Section 4. The 'first-

R. Atkinson

Scheme 1 Tropospheric reactions of alkoxyl radicals derived from alkanes. This example shows the 2-pentoxyl radical derived from n-pentane

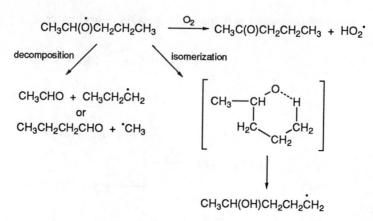

Scheme 2 Further reactions of the δ-hydroxyalkyl radicals formed in Scheme 1

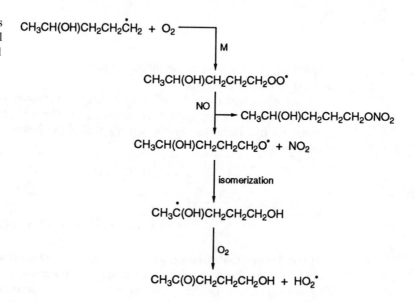

'generation' carbonyl products react further in the troposphere, as discussed in Section 6.

The methoxyl and ethoxyl radicals formed from methane and ethane, respectively, react only with O_2 in the troposphere, forming formaldehyde ($HCHO$) and acetaldehyde (CH_3CHO) respectively, as well as the $HO_2^{\bullet}$ radical:[5]

$$CH_3O^{\bullet} + O_2 \rightarrow HCHO + HO_2^{\bullet} \tag{19}$$

$$CH_3CH_2O^{\bullet} + O_2 \rightarrow CH_3CHO + HO_2^{\bullet} \tag{20}$$

If there is sufficient NO that the $HO_2^{\bullet}$ radical reaction with NO (reaction 21) dominates over the self-reaction of $HO_2^{\bullet}$ radicals or the reaction of the $HO_2^{\bullet}$ radical with O_3 (reactions 22 and 23),

$$HO_2^{\bullet} + NO \rightarrow {}^{\bullet}OH + NO_2 \tag{21}$$

74

$$HO_2^{\bullet} + HO_2^{\bullet} \xrightarrow{M} H_2O_2 + O_2 \qquad (22)$$

$$HO_2^{\bullet} + O_3 \rightarrow {}^{\bullet}OH + 2O_2 \qquad (23)$$

then efficient regeneration of the ${}^{\bullet}OH$ radical occurs and photolysis of the co-product NO_2 (reactions 6 and 3, page 69) leads to net formation of ozone.[10,19]

Presently, it is not possible to reliably predict the dominant reaction pathway(s) of most alkoxyl radicals in the troposphere, although empirical methods to calculate the relative importance of reaction with O_2 *versus* decomposition have been proposed.[24] It is only recently that direct evidence for alkoxyl radical isomerization at room temperature has become available,[48-50] with the studies of Atkinson and Aschmann[48,50] involving alkoxyl radicals formed from selected ketones[48] and alcohols.[50]

A tropospheric reaction scheme in the presence of NO is shown in Scheme 3 for the 2-butyl radical formed from n-butane (the first-generation products are italicized).

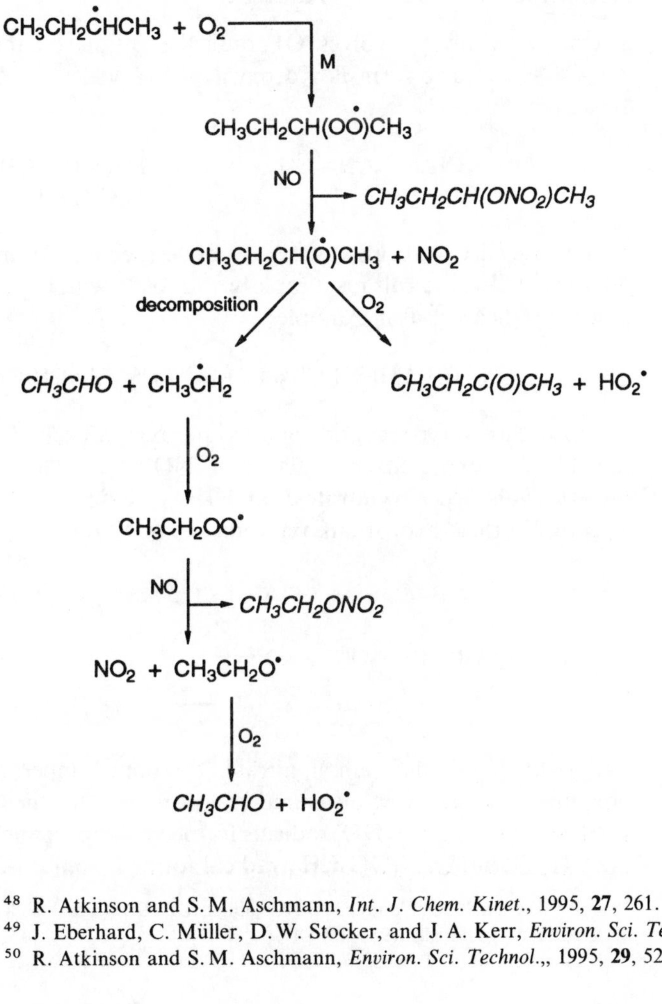

Scheme 3 Tropospheric reactions of n-butane (via the 2-butyl radical) in the presence of NO

$$CH_3CH_2\overset{\bullet}{C}HCH_3 + O_2 \xrightarrow{}$$

$$\downarrow M$$

$$CH_3CH_2CH(\overset{\bullet}{O}O)CH_3$$

$$\xrightarrow{NO} \quad CH_3CH_2CH(ONO_2)CH_3$$

$$CH_3CH_2CH(\overset{\bullet}{O})CH_3 + NO_2$$

decomposition $\swarrow$ $\qquad$ $\searrow$ O_2

$$CH_3CHO + CH_3\overset{\bullet}{C}H_2 \qquad\qquad CH_3CH_2C(O)CH_3 + HO_2^{\bullet}$$

$$\downarrow O_2$$

$$CH_3CH_2OO^{\bullet}$$

$$\xrightarrow{NO} \quad CH_3CH_2ONO_2$$

$$NO_2 + CH_3CH_2O^{\bullet}$$

$$\downarrow O_2$$

$$CH_3CHO + HO_2^{\bullet}$$

[48] R. Atkinson and S. M. Aschmann, *Int. J. Chem. Kinet.*, 1995, **27**, 261.
[49] J. Eberhard, C. Müller, D. W. Stocker, and J. A. Kerr, *Environ. Sci. Technol.*, 1995, **29**, 232.
[50] R. Atkinson and S. M. Aschmann, *Environ. Sci. Technol.*,, 1995, **29**, 528.

R. Atkinson

4 Tropospheric Chemistry of Alkenes

As noted in the Introduction, alkenes are emitted into the troposphere from anthropogenic sources (mainly combustion sources such as vehicle exhaust); and isoprene (2-methyl-1,3-butadiene), $C_{10}H_{16}$ monoterpenes, and $C_{15}H_{24}$ sesquiterpenes are emitted from vegetation. In the troposphere, alkenes react with ·OH radicals, NO_3· radicals, and O_3, and all three of these reactions must be considered in assessing the transformation processes of a given alkene (Table 1, page 71). All three of these reactions proceed wholly (NO_3· radical and O_3 reactions) or mainly (·OH radical reaction) by initial addition to the $>C=C<$ bond(s).[24,40–42] The ·OH radical reactions also involve H-atom abstraction from the C–H bonds of the $-CH_3$, $-CH_2-$, and $>CH-$ groups within the alkyl side-chains around the $>C=C<$ bond(s). For example, this H-atom abstraction pathway is calculated to account for ~10–15% of the overall ·OH radical reaction with 1-heptene.[24]

Reaction with the ·OH Radical

The major pathway involves ·OH radical addition to either carbon atom of the $>C=C<$ bond to form β-hydroxyalkyl radicals,[24] as shown for example for 1-butene:

$$\cdot OH + CH_3CH_2CH{=}CH_2 \xrightarrow{M} CH_3CH_2\dot{C}HCH_2OH$$
$$\text{and } CH_3CH_2CH(OH)\dot{C}H_2 \qquad (24)$$

As for the alkyl radicals discussed above (Section 3), in the troposphere the β-hydroxyalkyl radicals react rapidly and solely with O_2 to form β-hydroxyalkyl peroxyl radicals.[24] For example:

$$CH_3CH_2CH(OH)\dot{C}H_2 + O_2 \xrightarrow{M} CH_3CH_2CH(OH)CH_2OO\cdot \qquad (25)$$

Analogous to the reactions of alkyl peroxyl radicals, β-hydroxyalkyl peroxyl radicals are expected to react with NO, NO_2 (to form thermally labile β-hydroxyalkyl peroxynitrates), and HO_2· radicals.[24] The reaction with NO forms either the β-hydroxyalkoxyl radical plus NO_2 or the β-hydroxyalkyl nitrate:

$$CH_3CH_2CH(OH)CH_2OO\cdot + NO \longrightarrow$$
$$\xrightarrow{M} CH_3CH_2CH(OH)CH_2ONO_2 \qquad (26a)$$
$$\longrightarrow CH_3CH_2CH(OH)CH_2O\cdot + NO_2 \qquad (26b)$$

The yield of β-hydroxyalkyl nitrate, at room temperature and atmospheric pressure of air, has been measured as ~1.5–1.7% for the $CH_3CH(O\dot{O})CH_2OH$ and $CH_3CH(OH)CH_2OO\cdot$ radicals formed from propene[51] and 3.7 ± 0.9% for the $CH_3CH(OH)CH(O\dot{O})CH_3$ radical formed from cis-2-butene.[52] The HO_2·

[51] P. B. Shepson, E. O. Edney, T. E. Kleindienst, J. H. Pittman, G. R. Namie, and L. T. Cupitt, Environ. Sci. Technol., 1985, 19, 849.
[52] K. Muthuramu, P. B. Shepson, and J. M. O'Brien, Environ. Sci. Technol., 1993, 27, 1117.

radical reactions are expected to form β-hydroxyalkyl hydroperoxides;[24] for example:

$$CH_3CH_2CH(OH)CH_2OO^{\bullet} + HO_2^{\bullet} \rightarrow CH_3CH_2CH(OH)CH_2OOH + O_2 \quad (27)$$

As for the alkoxyl radicals formed from the alkanes, the β-hydroxyalkoxyl radicals react with O_2, unimolecularly decompose, or isomerize via a 1,5-H shift through a six-membered transition state.[24] For example, Scheme 4 shows the possible reactions of the $CH_3CH_2CH(OH)CH_2O^{\bullet}$ radical formed after $^{\bullet}OH$ radical addition to the internal carbon atom of the $>C{=}C<$ bond in 1-butene.

Scheme 4 Tropospheric reactions of β-hydroxy-alkoxyl radicals derived from alkenes (1-butene, here)

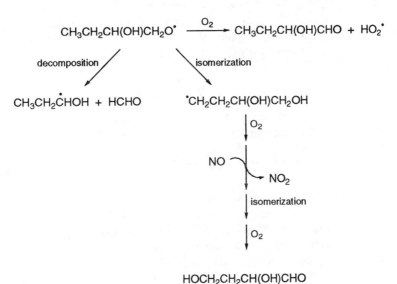

At room temperature and atmospheric pressure of air, the $HOCH_2CH_2O^{\bullet}$ radical formed after $^{\bullet}OH$ radical addition to ethene reacts with O_2 and thermally decomposes,[24] while the β-hydroxyalkoxyl radicals formed after $^{\bullet}OH$ radical addition to propene, 1-butene, and *trans*-2-butene primarily decompose to give shorter-chain α-hydroxyalkyl radicals ($R\dot{C}HOH$).[24]

C_1–C_4 α-hydroxyalkyl radicals such as the $^{\bullet}CH_2OH$ and $CH_3\dot{C}HOH$ radicals react rapidly with O_2, by initial addition, to form the corresponding carbonyl compound plus the $HO_2^{\bullet}$ radical.[24] For the $CH_3CH_2\dot{C}HOH$ radical formed after decomposition of the $CH_3CH_2CH(OH)CH_2O^{\bullet}$ radical (Scheme 4), the reaction with O_2 leads to the formation of propanal:

$$CH_3CH_2\dot{C}HOH + O_2 \rightarrow CH_3CH_2CHO + HO_2^{\bullet} \quad (28)$$

Therefore, for propene, 1-butene, and the 2-butenes, the products formed in the presence of NO are the same irrespective of which carbon atom the $^{\bullet}OH$ radical initially adds to: *they are the carbonyls arising from oxidative cleavage of the $>C{=}C<$ bond.*[24] For example, CH_3CHO plus $HCHO$ are formed from propene, CH_3CH_2CHO plus HCHO are formed from 1-butene, and two molecules of CH_3CHO are formed from the 2-butenes.[24]

However, it appears that for the more complex β-hydroxyalkoxyl radicals formed from the $\geq C_5$ 1-alkenes, isoprene, and $C_{10}H_{16}$ monoterpenes,[24,53] reaction with O_2 or, more likely, isomerization must occur to a significant, if not dominant, extent.[53] For example, for the reaction of the $^{\cdot}$OH radical with 1-octene in the presence of NO, the heptanal formation yield at room temperature and atmospheric pressure of air is only 15–21%.[53,54] Because H-atom abstraction from the C–H bonds of the C_6H_{13} alkyl group in 1-octene is estimated to account for only $\sim$15–20% of the overall reaction,[24] and formation of the β-hydroxyalkyl nitrate from the $RO_2^{\cdot}$ + NO reaction is expected to account for only a minor fraction of the overall reaction products, a large fraction of the β-hydroxyalkoxyl radicals formed in the $^{\cdot}$OH radical addition to 1-octene must react with O_2 or isomerize.

Reaction with the $NO_3^{\cdot}$ Radical

The reactions of the $NO_3^{\cdot}$ radical with alkenes proceed essentially entirely by initial addition of the $NO_3^{\cdot}$ radical to form a β-nitratoalkyl radical:[24,41]

$$NO_3^{\cdot} + CH_3CH{=}CH_2 \xrightarrow{M} CH_3\dot{C}HCH_2ONO_2 \text{ and } CH_3CH(ONO_2)\dot{C}H_2 \quad (29)$$

The subsequent reactions of the β-nitratoalkyl radicals are analogous to those of the β-hydroxyalkyl radicals formed from the corresponding $^{\cdot}$OH radical reactions,[24,41] except that if $NO_3^{\cdot}$ radicals are present in the troposphere, then NO will be at low concentrations due to the rapid reaction of NO with the $NO_3^{\cdot}$ radical.[5] Hence, β-nitratoalkyl peroxyl radicals are expected to react primarily with NO_2 [to form thermally unstable peroxynitrates such as $CH_3CH(OONO_2)CH_2ONO_2$] and $HO_2^{\cdot}$ radicals [with self-reactions or combination reactions with other peroxyl radicals being important in laboratory studies].[24,41]

Under laboratory conditions the reaction of propene with the $NO_3^{\cdot}$ radical leads to the formation of HCHO, CH_3CHO, and $CH_3C(O)CH_2ONO_2$.[24,41,51]

Reaction with O_3

O_3 initially adds to the $>C{=}C<$ bond to form an energy-rich primary ozonide, which rapidly decomposes (as shown in Scheme 5)[24] to form two sets of (carbonyl + biradical), where [ ]‡ denotes an energy-rich species. Recent studies have shown that the relative importance of the two decomposition pathways of the primary ozonide to form the two (carbonyl plus biradical) products depends on the structure of the alkene.[55–57] It appears that the two decomposition

[53] R. Atkinson, E. C. Tuazon, and S. M. Aschmann, *Environ. Sci. Technol.*, 1995, **29**, 1674.

[54] S. E. Paulson and J. H. Seinfeld, *Environ. Sci. Technol.*, 1992, **26**, 1165.

[55] O. Horie and G. K. Moortgat, *Atmos. Environ.*, 1991, **25A**, 1881.

[56] R. Atkinson and S. M. Aschmann, *Environ. Sci. Technol.*, 1993, **27**, 1357.

Scheme 5 Tropospheric reaction of alkenes with ozone

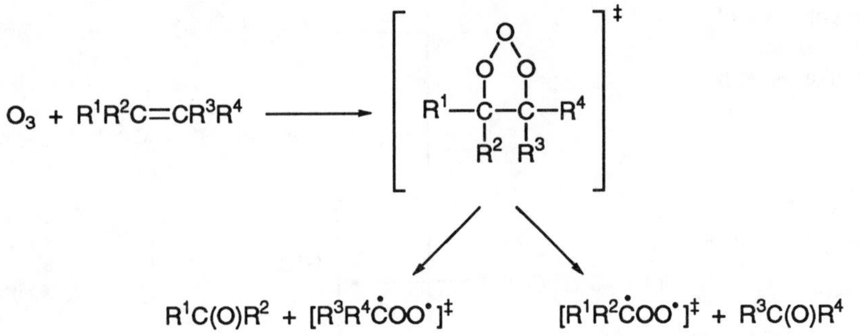

$$O_3 + R^1R^2C{=}CR^3R^4 \longrightarrow \left[\begin{array}{c} \overset{\displaystyle O}{\underset{\displaystyle}{\overset{\displaystyle \diagup\diagdown}{O\quad O}}} \\ R^1{-}\underset{\displaystyle R^2}{\overset{\displaystyle |}{C}}{-}\underset{\displaystyle R^3}{\overset{\displaystyle |}{C}}{-}R^4 \end{array}\right]^{\ddagger}$$

$$R^1C(O)R^2 + [R^3R^4\overset{\centerdot}{C}OO^{\centerdot}]^{\ddagger} \qquad\qquad [R^1R^2\overset{\centerdot}{C}OO^{\centerdot}]^{\ddagger} + R^3C(O)R^4$$

pathways of the primary ozonide are of approximately equal importance for alkenes of structure $RCH{=}CH_2$, $R^1CH{=}CHR^2$, and $R^1R^2C{=}CR^3R^4$,[56,57] but that for alkenes of structure $R^1R^2C{=}CH_2$ and $R^1R^2C{=}CHR^3$ the primary ozonide decomposes preferentially to form the dialkyl-substituted biradical $[R^1R^2\overset{\centerdot}{C}OO^{\centerdot}]^{\ddagger}$ plus $HCHO$ or R^3CHO.[56,57]

The fate of the initially energy-rich biradicals is presently not well understood.[24,55,57,58] These biradicals can be collisonally stabilized or decompose by a number of pathways,[24,57,58] as shown in Scheme 6. As expected, the stabilization pathway is pressure dependent, with the fraction of the overall biradical reaction being stabilized increasing with increasing pressure.[59] $O(^3P)$ atom elimination has not been observed for the acyclic alkenes at room temperature and atmospheric pressure,[60–63] but may be a minor pathway for certain of the cycloalkenes such as 1-methylcyclohexene and α-pinene.[57] In contrast, $^{\centerdot}OH$ radicals are formed from the reactions of O_3 with alkenes, often in unit or close to unit yield.[24,54,56–58,61] For acyclic alkenes, and cycloalkenes not containing terminal ${=}CH_2$ groups, $^{\centerdot}OH$ radical production increases with the number of alkyl substituent groups on the double bond or, approximately equivalently, with the number of alkyl substituents on the biradicals.[56] This formation of $^{\centerdot}OH$ radicals from the reactions of O_3 with alkenes leads to secondary reactions of the $^{\centerdot}OH$ radical with the alkenes. Therefore, unless $^{\centerdot}OH$ radicals are scavenged, the reactions of O_3 with alkenes also involve $^{\centerdot}OH$ radicals, and the products observed and their yields may not be those for the O_3 reactions.[64]

A number of studies have shown the formation of H_2O_2 and organic

[57] R. Atkinson, E. C. Tuazon, and S. M. Aschmann, *Environ. Sci. Technol.*, 1995, **29**, 1860.
[58] O. Horie, P. Neeb, and G. K. Moortgat, *Int. J. Chem. Kinet.*, 1994, **26**, 1075.
[59] S. Hatakeyama, H. Kobayashi, and H. Akimoto, *J. Phys. Chem.*, 1984, **88**, 4736.
[60] H. Niki, P. D. Maker, C. M. Savage, L. P. Breitenbach, and R. I. Martinez, *J. Phys. Chem.*, 1984, **88**, 766.
[61] H. Niki, P. D. Maker, C. M. Savage, L. P. Breitenbach, and M. D. Hurley, *J. Phys. Chem.*, 1987, **91**, 941.
[62] R. Atkinson, J. Arey, S. M. Aschmann, and E. C. Tuazon, *Res. Chem. Intermed.*, 1994, **20**, 385.
[63] R. Atkinson, S. M. Aschmann, J. Arey, and E. C. Tuazon, *Int. J. Chem. Kinet.*, 1994, **26**, 945.
[64] H. Hakola, J. Arey, S. M. Aschmann, and R. Atkinson, *J. Atmos. Chem.*, 1994, **18**, 75.

R. Atkinson

Scheme 6 Possible fates of biradicals formed in the reaction of alkenes with ozone

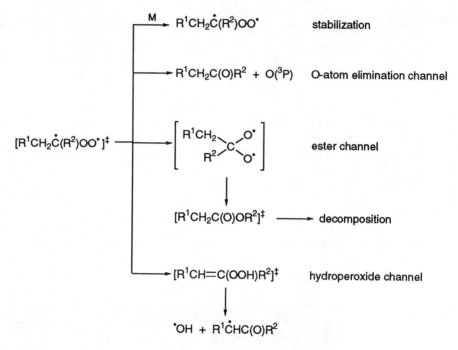

hydroperoxides from the reactions of O_3 with alkenes.[65-70] While there are significant discrepancies between the reported H_2O_2 formation yields, the most recent studies[69,70] indicate that the H_2O_2 yields increase with the concentration of water vapour and are $\sim$0.1–3% at room temperature and 10–50% relative humidity. The formation of H_2O_2 and hydroxymethyl hydroperoxide ($HOCH_2OOH$) have been explained by the reaction of the stabilized $^{\bullet}CH_2OO^{\bullet}$ biradical with water vapour:[69,70]

$$^{\bullet}CH_2OO^{\bullet} + H_2O \longrightarrow H_2O_2 + HCHO \quad (30a)$$

$$\longrightarrow [HOCH_2OOH] \longrightarrow HC(O)OH + H_2O \quad (30b)$$

$$\downarrow M$$

$$HOCH_2OOH$$

[65] S. Gäb, E. Hellpointner, W. V. Turner, and F. Kórte, *Nature (London)*, 1985, **316**, 535.
[66] K. H. Becker, K. J. Brockmann, and J. Bechara, *Nature (London)*, 1990, **346**, 256.
[67] C. N. Hewitt and G. L. Kok, *J. Atmos. Chem.*, 1991, **12**, 181.
[68] R. Simonaitis, K. J. Olszyna, and J. F. Meagher, *Geophys. Res. Lett.*, 1991, **18**, 9.
[69] K. H. Becker, J. Bechara, and K. J. Brockmann, *Atmos. Environ.*, 1993, **27A**, 57.
[70] S. Hatakeyama, H. Lai, S. Gao, and K. Murano, *Chem. Lett.*, 1993, 1287.

5 Tropospheric Chemistry of Aromatic Compounds

Aromatic Hydrocarbons

Benzene and the alkyl-substituted benzenes such as toluene, ethylbenzene, the xylenes, and the trimethylbenzenes react with ˙OH radicals and NO₃˙ radicals,[24,40,41] with the ˙OH radical reactions dominating as the tropospheric removal/transformation process (Table 1, page 71).

The ˙OH radical reactions proceed by H-atom abstraction from the C–H bonds of the alkyl substituent groups (or in the case of benzene from the C–H bonds of the aromatic ring), and by ˙OH radical addition to the aromatic ring to form a hydroxycyclohexadienyl, or alkyl-substituted hydroxycyclohexadienyl, radical (hereafter termed an ˙OH–aromatic adduct).[24,40] These two reactions are illustrated in Scheme 7, using toluene as an example.

Scheme 7 Reactions of the ˙OH radical with toluene

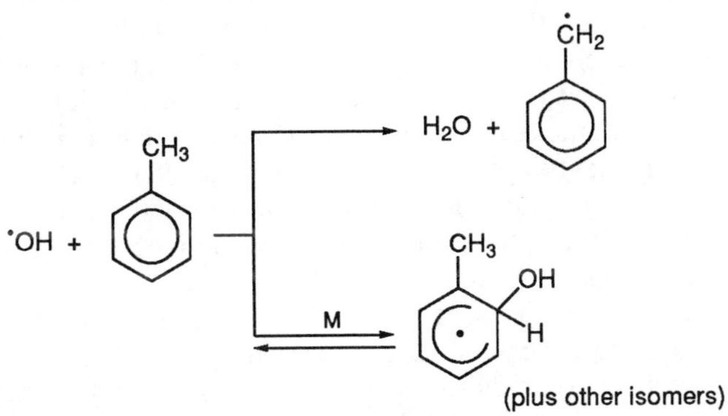

(plus other isomers)

The ˙OH radical addition pathway is reversible because of thermal decomposition of the ˙OH–aromatic adduct,[24,40] and the lifetimes of the ˙OH–benzene and ˙OH–toluene adducts due to thermal decomposition are each ∼0.2–0.3 s at 298 K and ∼0.025 s at 325 K.[24,71]

The H-atom abstraction pathway accounts for ≤10% of the overall ˙OH radical reactions with benzene and the methyl-substituted benzenes at room temperature and atmospheric pressure.[24,40] The benzyl and alkyl-substituted benzyl radicals behave in the troposphere similarly to alkyl radicals (and can be considered phenyl-substituted alkyl radicals),[24] and therefore their tropospheric degradation schemes are similar. For example, the reactions for the benzyl radical, PhĊH₂, in the presence of NO are

$$PhĊH_2 + O_2 \xrightarrow{M} PhCH_2OO˙ \tag{31}$$

$$PhCH_2OO˙ + NO \longrightarrow \begin{cases} \xrightarrow{M} PhCH_2ONO_2 \quad \text{(benzyl nitrate)} & \text{(32a)} \\ \\ \longrightarrow PhCH_2O˙ + NO_2 & \text{(32b)} \end{cases}$$

[71] R. Knispel, R. Koch, M. Siese, and C. Zetzsch, *Ber. Bunsen-Ges. Phys. Chem.*, 1990, **94**, 1375.

$$PhCH_2O^{\cdot} + O_2 \rightarrow PhCHO + HO_2^{\cdot} \qquad (33)$$

The benzyl nitrate yield from the reaction of NO with the benzyl peroxyl radical is ~ 10–12% at room temperature and atmospheric pressure.[24]

The major reaction pathway involves the formation of the ˙OH–aromatic adduct, but only in the past few years has the fate of the ˙OH–aromatic adducts become known.[24,71,72] Kinetic studies of the reactions of the ˙OH–benzene and ˙OH–toluene adducts with NO, NO$_2$, and O$_2$ have been carried out by several groups,[24] and no reaction with NO has been observed.[71] However, reactions of the ˙OH–benzene and ˙OH–toluene adducts with O$_2$ and NO$_2$ are observed,[24,71] with room temperature rate constants of 1.8–5.4×10^{-16} cm^3 molecule^{-1} s^{-1} and 2.5–3.6×10^{-11} cm^3 molecule^{-1} s^{-1}, respectively.[71] These kinetic data indicate that in the lower troposphere, including in polluted urban air masses, the dominant reaction of the ˙OH–benzene and ˙OH–toluene adducts (and, by analogy, the ˙OH–alkyl-substituted-benzene adducts) is with O$_2$ [the O$_2$ and NO$_2$ reactions are of equal importance for NO$_2$ mixing ratios of 1.5×10^{-6} and 3×10^{-6} for the ˙OH–benzene and ˙OH–toluene adducts, respectively[71]].

Product studies of the reactions of the ˙OH radical with toluene and *o*-xylene are consistent with these kinetic data for the ˙OH–aromatic adducts.[24,72] Interestingly, *o*-cresol is observed from the reactions of the ˙OH–toluene adduct with both O$_2$ and NO$_2$, and 2,3-butanedione is observed from the reactions of the ˙OH–*o*-xylene adducts with both O$_2$ and NO$_2$.[72] The formation yield of 2,3-butanedione from the ˙OH radical-initiated reaction of *o*-xylene is 18.5% when the ˙OH–*o*-xylene adducts react with O$_2$ (in the presence of sufficient NO to convert peroxyl radicals to alkoxyl radicals), and 4% when the ˙OH–*o*-xylene adducts react with NO$_2$.[72]

There is a need to determine the mechanisms and the products formed from the reactions of ˙OH–aromatic adducts with O$_2$ and NO$_2$. While the ˙OH–aromatic adducts, at least for benzene, toluene, the xylenes, and the trimethylbenzenes, will react with O$_2$ in the troposphere, it should be recognized that the literature product data[24] have generally been obtained from studies carried out with NO$_2$ concentrations markedly higher than those encountered in the troposphere. These laboratory product studies may therefore have involved reactions of the ˙OH–aromatic adducts with both O$_2$ and NO$_2$, and care should be exercised in using these product, and product yield, data to formulate detailed chemical mechanisms for the tropospheric photo-oxidations of aromatic hydrocarbons.

While the mechanisms of the reactions of the ˙OH–aromatic adducts with O$_2$ and NO$_2$ are not presently understood, product data and product formation yields are available for the ˙OH radical-initiated reactions of benzene, toluene, and *o*-, *m*-, and *p*-xylene[24,72] (recognizing that most of these data were obtained in the presence of NO$_x$ and hence may not be applicable to solely the O$_2$ reaction or solely the NO$_2$ reaction with the ˙OH–aromatic adducts). Taking toluene as an example, the products arising from the ˙OH radical addition pathway are the ring-retaining products *o*-, *m*-, and *p*-cresol (mainly the *ortho* isomer), and nitrotoluenes (obviously formed only from the reaction of NO$_2$ with the ˙OH–toluene adduct), together with the ring-cleavage products glyoxal

[72] R. Atkinson and S. M. Aschmann, *Int. J. Chem. Kinet.*, 1994, **26**, 929.

[(CHO)$_2$] and methylglyoxal (CH$_3$C(O)CHO).[24,72] The co-products to the α-dicarbonyls glyoxal and methylglyoxal may be 1,4-unsaturated carbonyls such as CH$_3$C(O)CH=CHCHO and CHOCH=CHCHO, respectively, although this has not yet been quantitatively confirmed. Furthermore, a variety of unsaturated carbonyl, dicarbonyl, and hydroxycarbonyl compounds have been observed.[24]

The products observed and quantified to date from the ˙OH radical-initiated reactions of benzene, toluene, and the xylenes account for only ~ 30–50% of the carbon reacted.[24]

Analogous to the reactions of the ˙OH radical with the aromatic hydrocarbons, the NO$_3$˙ radical reactions proceed by H-atom abstraction from the C–H bonds of the alkyl substituent groups, and by reversible addition to the aromatic rings to form an NO$_3$˙–aromatic adduct.[24,41] Because of the rapid thermal decomposition of the NO$_3$˙–monocyclic aromatic adducts,[24,41] with an estimated lifetime[41] at 298 K due to thermal decomposition of ~ 10^{-8} s, addition of the NO$_3$˙ radical to the aromatic ring is of no importance in the troposphere for monocyclic aromatic hydrocarbons.[24,41] The observed slow reactions of the alkyl-substituted benzenes therefore proceed by H-atom abstraction from the C–H bonds of the alkyl groups, and this is confirmed by the observation of a significant deuterium isotope effect for toluene, *o*-xylene, and *p*-xylene.[24,41]

Styrenes

Styrene [C$_6$H$_5$CH=CH$_2$] and its methyl-substituted homologues react with ˙OH radicals, NO$_3$˙ radicals, and O$_3$ (Table 1, page 71).[24] The magnitude of the rate constants and the products formed from these reactions[24] show that these reactions proceed by initial addition at the >C=C< bond in the substituent group, and styrene behaves as a phenyl-substituted ethene.[24] Accordingly, the reaction mechanisms and the products formed are analogous to those for the corresponding reactions of alkenes (Section 4).

Aromatic Aldehydes

Kinetic data are available only for benzaldehyde.[24] Based on the measured rate constants for the reactions of benzaldehyde with ˙OH radicals and NO$_3$˙ radicals,[24,40,41] benzaldehyde behaves similarly to an aliphatic aldehyde (Section 6).

Phenolic Compounds

Phenol, the cresols, and the dimethylphenols have been reported as products of the ˙OH radical-initiated reactions of benzene, toluene, and the xylenes, respectively.[24] Phenol, the cresols, and the dimethylphenols react with ˙OH radicals, NO$_3$˙ radicals, and O$_3$,[24] with the reactions of the cresols with O$_3$ being slow.[24] The major tropospheric transformation processes are therefore reaction with ˙OH radicals and NO$_3$˙ radicals (Table 1, page 71). The ˙OH radical reactions are analogous to the reactions of the ˙OH radical with aromatic hydrocarbons in that the reactions proceed by H-atom abstraction from the C–H and O–H bonds of the substituent –OH and alkyl groups, and by 'OH radical addition to the aromatic ring.[24,40] The H-atom abstraction pathway is of minor importance at room temperature, accounting for $\leqslant 10\%$ of the overall reaction for phenol and o-cresol.[24,40]

In the presence of NO_x, the phenoxyl (or alkyl-substituted phenoxyl) radical formed after H-atom abstraction from the –OH group appears to react with NO_2 to form 2-nitrophenol or alkyl-substituted 2-nitrophenols.[24,41] The reaction of the 'OH radical with phenol in the presence of NO_x is shown in Scheme 8.

Scheme 8 Reactions of the 'OH radical with phenol

(plus other isomers)

$+ H_2O$

NO_2

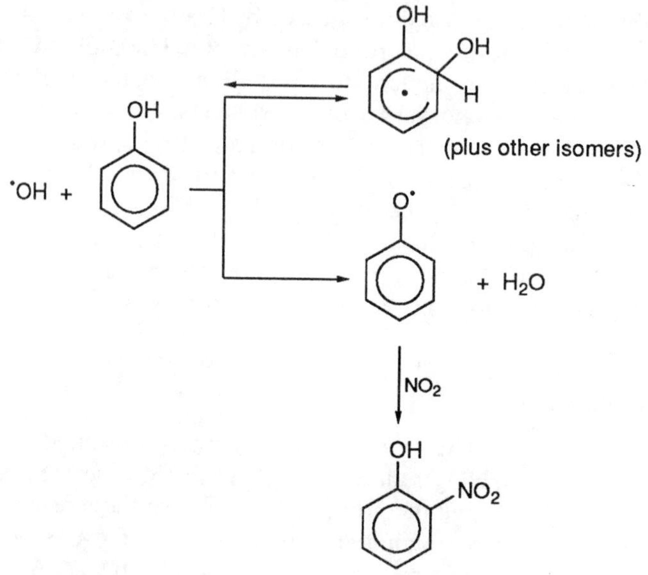

The products and mechanisms of the reactions of the 'OH–phenol and 'OH–cresol adducts are incompletely understood, although it is known that in the troposphere the sole reaction of the 'OH–phenol adduct is with O_2.[24]

The reactions of the NO_3' radical with phenolic compounds are rapid,[24,41] and lead to nitrophenol and alkyl-substituted nitrophenol formation in significant, but not unit, yield.[24] Taking phenol as an example, the available product data suggest that these NO_3' radical reactions proceed as shown in Scheme 9.

6 Tropospheric Chemistry of Oxygen-containing Compounds

In this section, the tropospheric chemistry of the major oxygen-containing compounds emitted into the troposphere, or formed *in situ* in the troposphere as a result of atmospheric transformations, is briefly discussed. The classes of oxygen-containing compounds considered are aliphatic aldehydes, ketones, and α-dicarbonyls (many of which are formed in the troposphere), alcohols, ethers, and α,β-unsaturated carbonyl compounds.

Scheme 9 Reactions of the NO$_3$ radical with phenol

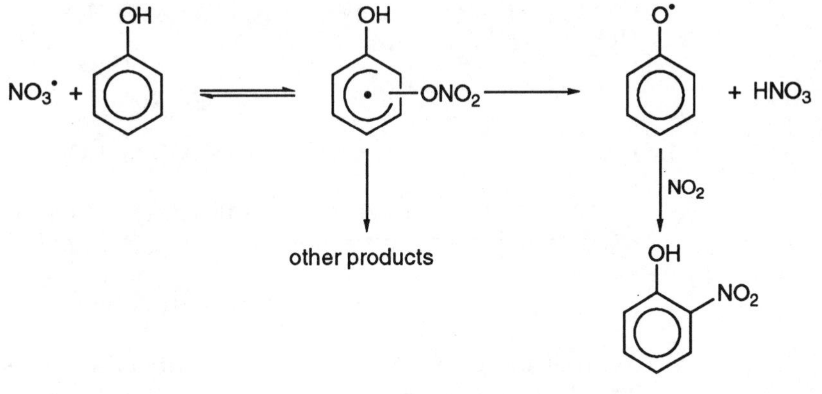

Aliphatic Aldehydes, Ketones, and α-Dicarbonyls

As shown in Table 1 (page 71), the major tropospheric transformation processes for the aliphatic aldehydes (including benzaldehyde), ketones, and α-dicarbonyls are photolysis and reaction with the ˙OH radical. Reactions with O$_3$ have not been observed at room temperature, and the NO$_3$ radical and HO$_2$ radical reactions are of negligible or minor importance in the troposphere.[5,24,41,42] Unfortunately, few data are available concerning the photolysis quantum yields for carbonyl compounds other than formaldehyde, acetaldehyde, propanal, and acetone.[5,24] Photolysis is calculated to be the dominant tropospheric loss process for formaldehyde and the three α-dicarbonyls studied to date (glyoxal, methylglyoxal, and 2,3-butanedione). Photolysis is calculated to be competitive with the ˙OH radical reaction as a tropospheric loss process for acetone (Table 1), while the ˙OH radical reactions are expected to be the dominant loss process for the higher aldehydes and ketones.

Photolysis of formaldehyde and acetaldehyde proceed by the following reactions:

$$HCHO + h\nu \longrightarrow H_2 + CO \tag{34a}$$
$$\longrightarrow H^{\bullet} + H\dot{C}O \tag{34b}$$

$$CH_3CHO + h\nu \longrightarrow CH_4 + CO \tag{35a}$$
$$\longrightarrow {}^{\bullet}CH_3 + H\dot{C}O \tag{35b}$$

The reactions of the ˙CH$_3$ radical, leading to HCHO, have been dealt with in Section 3 (see equations 12, 13b, and 19). In the troposphere, H˙ atoms and H$\dot{C}$O (formyl) radicals react only with O_2 to form the $HO_2^{\bullet}$ radical, and CO plus the $HO_2^{\bullet}$ radical, respectively:[5,24]

$$H^{\bullet} + O_2 \xrightarrow{M} HO_2^{\bullet} \tag{36}$$

$$H\dot{C}O + O_2 \rightarrow HO_2^{\bullet} + CO \tag{37}$$

The reactions of the $^{\bullet}OH$ radical with aldehydes proceed mainly (or totally for HCHO) by H-atom abstraction from the –CHO group.[24,40]

$$^{\bullet}OH + RCHO \rightarrow H_2O + R\dot{C}O \tag{38}$$

As noted above, the $H\dot{C}O$ radical reacts with O_2 by an abstraction pathway to form $HO_2^{\bullet} + CO$. In contrast, $R\dot{C}O$ (acyl) radicals from equation 38 with R = alkyl react in the troposphere with O_2 by addition to form an acyl peroxyl ($RC(O)OO^{\bullet}$) radical,[5,24]

$$CH_3\dot{C}O + O_2 \xrightarrow{M} CH_3C(O)OO^{\bullet} \tag{39}$$

and an analogous reaction occurs for the benzoyl ($C_6H_5\dot{C}O$) radical.[24] In the troposphere, acyl peroxyl radicals react with NO, NO_2, and $HO_2^{\bullet}$ radicals,[5,24] similarly to the alkyl peroxyl radicals dealt with in Section 3. For example:

$$CH_3C(O)OO^{\bullet} + NO \longrightarrow CH_3C(O)O^{\bullet} + NO_2 \tag{40}$$

$$\downarrow$$

$$\dot{C}H_3 + CO_2$$

$$CH_3C(O)OO^{\bullet} + NO_2 \xrightarrow{M} CH_3C(O)OONO_2 \text{ (peroxyacetyl nitrate; PAN)} \tag{41}$$

$$CH_3C(O)OO^{\bullet} + HO_2^{\bullet} \longrightarrow CH_3C(O)OH + O_3 \tag{42a}$$

$$\longrightarrow CH_3C(O)OOH + O_2 \tag{42b}$$

The reactions of PAN are dealt with in Section 7 below (equation 48). In the presence of NO, the $^{\bullet}OH$ radical-initiated reaction of acetaldehyde therefore leads (via the $^{\bullet}CH_3$, $CH_3OO^{\bullet}$, and $CH_3O^{\bullet}$ radicals, see Section 3) to the formation of HCHO; and the degradation reactions of higher aldehydes 'cascade' through the lower aldehydes to ultimately form HCHO (which then reacts to form CO and then CO_2).

The reactions of the $^{\bullet}OH$ radical with ketones are generally similar to the reactions of the $^{\bullet}OH$ radical with alkanes,[24] and the subsequent reactions of the substituted alkyl radicals produced in the initial reaction are analogous to those for the alkyl radicals discussed in Section 3 above.

Alcohols and Ethers

The dominant chemical loss process for the alcohols and ethers in the troposphere is by reaction with the $^{\bullet}$OH radical (Table 1, page 71). For the alcohols, the $^{\bullet}$OH radical reactions proceed by H-atom abstraction from the various C–H bonds and the O–H bond. For example, for ethanol:

$$^{\bullet}\text{OH} + \text{CH}_3\text{CH}_2\text{OH}$$

$$\longrightarrow \text{H}_2\text{O} + \text{CH}_3\text{CH}_2\text{O}^{\bullet} \quad (43a)$$

$$\longrightarrow \text{H}_2\text{O} + \text{CH}_3\dot{\text{C}}\text{HOH} \quad (43b)$$

$$\longrightarrow \text{H}_2\text{O} + {}^{\bullet}\text{CH}_2\text{CH}_2\text{OH} \quad (43c)$$

The radicals formed in reactions (43a), (43b), and (43c) react as discussed in Sections 3 and 4 to form acetaldehyde, acetaldehyde, and formaldehyde or glycolaldehyde (HOCH$_2$CHO), respectively.[5,24] [Note that the $^{\bullet}$CH$_2$CH$_2$OH radical formed in reaction (43c) is identical to that formed by $^{\bullet}$OH radical addition to ethene].

The reactions of the substituted alkyl radicals formed from the reactions of the $^{\bullet}$OH radical with the ethers are similar to the reactions of the alkyl radicals discussed in Section 3, and detailed discussion of the tropospheric degradation reactions of methyl t-butyl ether (MTBE) and ethyl t-butyl ether is given in the review of Atkinson.[24]

α,β-Unsaturated Carbonyl Compounds

This class of oxygen-containing compounds includes methacrolein [CH$_2$=C(CH$_3$)CHO] and methyl vinyl ketone [CH$_3$C(O)CH=CH$_2$], the tropospheric degradation products of isoprene.[24,73] The dominant tropospheric reaction of the α,β-unsaturated carbonyls is with the $^{\bullet}$OH radical, with the O$_3$ and NO$_3$$^{\bullet}$ radical reactions being of relatively minor importance.[24]

For the α,β-unsaturated aldehydes such as acrolein (CH$_2$=CHCHO), crotonaldehyde (CH$_3$CH=CHCHO), and methacrolein, the $^{\bullet}$OH radical (and NO$_3$$^{\bullet}$ radical) reactions proceed by H-atom abstraction from the C–H bond of the –CHO group and $^{\bullet}$OH radical addition to the >C=C< bond.[24,40,41] For example, for methacrolein:

$$^{\bullet}\text{OH} + \text{CH}_2\text{=C(CH}_3)\text{CHO}$$

$$\longrightarrow \text{H}_2\text{O} + \text{CH}_2\text{=C(CH}_3)\dot{\text{C}}\text{O} \quad (\sim 50\%) \ (44a)$$

$$\longrightarrow \text{HOCH}_2\dot{\text{C}}(\text{CH}_3)\text{CHO} \quad (\sim 50\%) \ (44b)$$
$$\text{or}$$
$$^{\bullet}\text{CH}_2\text{C(OH)(CH}_3)\text{CHO}$$

73 S. M. Aschmann and R. Atkinson, *Environ. Sci. Technol.*, 1994, **28**, 1539.

The subsequent reactions of the acyl radical $CH_2=C(CH_3)\dot{C}O$ and of the β-hydroxyalkyl radicals $HOCH_2\dot{C}(CH_3)CHO$ and $\dot{C}H_2C(OH)(CH_3)CHO$ are analogous to those of the acyl ($R\dot{C}O$) and β-hydroxyalkyl radicals discussed above and in Section 4. However, for the β-hydroxyalkoxyl radicals formed subsequent to $\dot{O}H$ radical addition to the $>C=C<$ bond of α,β-unsaturated carbonyls, the 'first-generation' products depend on which carbon atom the $\dot{O}H$ radical adds to. For example, for the $HOCH_2C(\dot{O})(CH_3)CHO$ and $\dot{O}CH_2C(OH)(CH_3)CHO$ alkoxyl radicals formed after $\dot{O}H$ radical addition to methacrolein:[24]

$$HOCH_2C(\dot{O})(CH_3)CHO \longrightarrow HOCH_2C(O)CH_3 + H\dot{C}O \qquad (45)$$

$$\dot{O}CH_2C(OH)(CH_3)CHO \longrightarrow HCHO + CH_3\dot{C}(OH)CHO \qquad (46)$$

$$\downarrow O_2$$

$$CH_3C(O)CHO + HO_2\dot{}$$

The reactions of the $\dot{O}H$ radical with α,β-unsaturated ketones proceed by initial addition of the $\dot{O}H$ radical to the $>C=C<$ bond, analogous to the corresponding pathway in the α,β-unsaturated aldehydes.[24] The subsequent reactions of the resulting β-hydroxyalkyl radicals are as discussed in Section 4, and the β-hydroxyalkoxyl radical reactions are analogous to reactions 45 and 46 for the methacrolein system.

7 Tropospheric Chemistry of Nitrogen-containing Compounds

The nitrogen-containing compounds of interest here are the alkyl nitrates and β-hydroxyalkyl nitrates formed from the reactions of the organic peroxyl radicals with NO, and the peroxynitrates formed from the reactions of NO_2 with alkyl and acyl peroxyl radicals. The only important reaction of the peroxynitrates in the lower troposphere is thermal decomposition:

$$ROONO_2 \xrightarrow{M} RO_2\dot{} + NO_2 \qquad (47)$$

$$RC(O)OONO_2 \xrightarrow{M} RC(O)OO\dot{} + NO_2 \qquad (48)$$

with thermal decomposition lifetimes at 298 K and atmospheric pressure of 0.1–1 s for the alkyl peroxynitrates $ROONO_2$,[5,24] and ~ 30 min for the acyl peroxynitrates $RC(O)OONO_2$.[5,24] Because the thermal decomposition rate constants decrease with decreasing temperature,[5,24] the thermal decomposition lifetimes increase markedly with increasing altitude in the troposphere; and in the upper troposphere photolysis and reaction with the $\dot{O}H$ radical may be important for peroxynitrates.

The alkyl nitrates, $RONO_2$, undergo photolysis and reaction with the $\dot{O}H$

radical in the troposphere.[5,24] Photolysis involves cleavage of the $RO-NO_2$ bond,[24] and the tropospheric lifetimes of the alkyl nitrates studied to date with respect to photolysis are in the range ~ 15–30 days.[24] The reactions of the ˙OH radical with the alkyl nitrates appear to involve H-atom abstraction from the C–H bonds, to form nitrato-substituted alkyl radicals.[24] It is expected that the subsequent reactions of these substituted alkyl radicals are similar to those for the alkyl radicals formed from the reaction of the ˙OH radical with alkanes (Section 3).

8 Conclusions

Organic compounds present in the gas phase in the troposphere are degraded by photolysis and/or chemical reaction with ˙OH radicals, NO_3˙ radicals, and O_3, with the organic compound being transformed to oxidized, and generally more polar, products which often have less carbon atoms than the parent compound. These degradation reactions continue until the original organic compound has been degraded to CO_2 and H_2O, or the products are removed from the atmosphere by wet deposition and/or dry deposition. In this degradation process, gas-to-particle conversion may occur, enhancing the possibility of wet and dry deposition.[3] Through these chemical and physical processes, the atmosphere is fortunately 'cleansed' of the organic species emitted into it.

Alternatives to CFCs and their Behaviour in the Atmosphere

PAULINE M. MIDGLEY

1 Introduction: Alternatives to CFCs

Why are Alternatives Needed?

The use of chlorofluorocarbons (CFCs) is being phased out under the provisions of the United Nations Environment Programme (UNEP) Montreal Protocol on Substances that Deplete the Ozone Layer. CFCs (fully-halogenated methanes and ethanes containing a combination of chlorine and fluorine) were one of the most widely applied classes of chemicals because of their desirable combination of chemical stability and physical properties. Following their introduction in the 1930s as safe refrigerants, applications of CFCs grew to encompass use as aerosol propellants, in air conditioning, as blowing agents for expanded foam insulation and packaging, and as cleaning solvents. As non-flammable chemicals with low toxicity and reactivity, CFCs were acceptable from the standpoint of worker and consumer safety. Annual production grew steadily, reaching a peak total of over 800 000 metric tonnes (1 metric tonne = 10^3 kg) of CFCs 11 and 12 ($CFCl_3$ and CF_2Cl_2) by 1974, the year in which their potential role in ozone depletion was first postulated.[1] The very stability that made CFCs suitable in so many applications also gave rise to this environmental concern and eventually resulted in their negotiated phase-out under the terms of the Montreal Protocol[2] and its subsequent revisions.

Unless they are consumed in a destruction or transformation process, CFCs eventually escape into the atmosphere from all applications because of their high vapour pressure. There is a characteristic delay before this release for each application, ranging from virtually instantaneous release from aerosol propellants to a delay of decades for the blowing agents captured in the bubbles of insulation foam.[3] Once emitted into the atmosphere, CFCs can persist for more than 100 years. There is no known destruction mechanism for CFCs in the troposphere and thus they are transported into the stratosphere, where they break down and release chlorine when exposed to solar UV radiation. The released chlorine components can take part in reactions that destroy ozone molecules.

[1] M.J. Molina and F.S. Rowland, *Nature (London)*, 1974, **249**, 810.
[2] 'Montreal Protocol on Substances that Deplete the Ozone Layer, Final Report', United Nations Environment Programme, New York, 1987.
[3] P.H. Gamlen, B.C. Lane, P.M. Midgley, and J.M. Steed, *Atmos. Environ.*, 1986, **20**, 1077.

The ozone layer lies in the stratosphere, typically 13–40 km above the Earth's surface, and acts as a protective blanket filtering out most of the harmful UV radiation from the Sun. Thus, decreases in stratospheric ozone can lead to increases in the level of UV at the ground with the potential for adverse effects on some plants and animals as well as human health. A considerable body of research has been undertaken in the last 20 years to refine understanding of the links between stratospheric chlorine and ozone depletion, and the subsequent effects. Results of this work have been extensively reviewed by the international scientific community.[4,5] Observations of a hole in the ozone layer over Antarctica occurring each spring[6] added urgency to the negotiations for a phase-out of CFCs and other long-lived, chlorine- and bromine-containing species. Moreover, measurements from the ground and from satellites show a long term loss of ozone from the layer over the Northern Hemisphere outside of the tropics. The reduction is between 3% and 6% per decade, depending on latitude and season.

The Montreal Protocol was first agreed upon in 1987 and has been ratified by over 100 countries. It has subsequently been revised twice in response to scientific developments. The latest amendments, adopted in Copenhagen in 1992, call for a complete phase-out of halons (bromine-containing halomethanes used as fire-fighting agents) by 1994 and of CFCs, carbon tetrachloride, and methyl chloroform (1,1,1-trichloroethane, CH_3CCl_3) by 1996 in developed countries, with a 10 year grace period for developing countries. It is now well established[7–9] that the *growth* rates of CFCs 11 and 12 in the atmosphere have already started to decline, reflecting the rapid switch into alternatives in many applications ahead of their phase-out. Most recently, there have been preliminary reports of an actual decline in atmospheric *abundances* of some of the controlled species.

Selection of Alternatives

In coming up with alternatives to meet the demand previously serviced by CFCs, while continuing to ensure worker and consumer safety, and at the same time reducing the risk to the environment, an integrated approach to product selection must be adopted. Factors to be considered include flammability, toxicity, and reactivity; physical properties determining suitability for the specific application; capital equipment and operating costs; and environmental concerns, both local and global.

There is no single solution to the replacement of CFCs in all applications. The technical options may be considered in four categories:[10] conservation, non-

[4] World Meteorological Organization, 'Scientific Assessment of Ozone Depletion: 1991', Global Ozone Research and Monitoring Project—Report no. 25, WMO, Geneva, 1991.

[5] World Meteorological Organization, 'Scientific Assessment of Ozone Depletion: 1994', Global Ozone Research and Monitoring Project—Report no. 37, WMO, Geneva, 1994.

[6] J.C. Farman, B.G. Gardiner, and J.D. Shanklin, *Nature (London)*, 1985, **315**, 207.

[7] J.W. Elkins, T.M. Thompson, T.H. Swanson, J.H. Butler, B.D. Hall, S.O. Cummings, D.A. Fisher, and A.G. Raffo, *Nature (London)*, 1993, **364**, 780.

[8] M.A.K. Khalil and R.A. Rasmussen, *J. Geophys. Res.*, 1993, **98**, 23091.

[9] D.M. Cunnold, P.J. Fraser, R.F. Weiss, R.G. Prinn, P.G. Simmonds, B.R. Miller, F.N. Alyea, and A.J. Crawford, *J. Geophys. Res.*, 1994, **99**, 1107.

fluorocarbon or not-in-kind (NIK) substitutes, hydrochlorofluorocarbons (HCFCs), and hydrofluorocarbons (HFCs). Conservation includes improved design of refrigeration and air conditioning equipment, and the recovery and recycling of refrigerants and cleaning solvents. Non-fluorocarbon substitutes include hydrocarbons used as aerosol propellants or foam-blowing agents, and aqueous cleaning systems. (The chemistry of these hydrocarbons in the atmosphere is discussed in Chapter 5, page 65). It has been estimated that about three-quarters of future demand can be met through increased conservation and the use of non-fluorocarbon alternatives.[11] The remaining quarter, predominantly the servicing of existing refrigeration and air conditioning equipment, requires the use of compounds with similar properties to the CFCs such as HCFCs and HFCs, in order to achieve the CFC phase-out targets and at the same time permit the equipment to be operated for its useful economic life.

Hydrochlorofluorocarbons (HCFCs) and Hydrofluorocarbons (HFCs)

As members of the fluorocarbon family, HCFCs and HFCs retain many of the desirable properties of CFCs but, because they contain hydrogen, they decompose in the lower atmosphere. This results in much shorter atmospheric lifetimes compared to the CFCs and reduces their potential contribution to ozone depletion (in the case of the chlorine-containing HCFCs) and to global warming. HCFCs play an important role as interim substitutes for CFCs, both reducing peak chlorine loading in the lower stratosphere and accelerating the recovery to the level prior to the first observation of an Antarctic ozone hole (2 ppb, or 2 parts in 10^9). However, because of their own ozone depletion potential, HCFCs are controlled under the Montreal Protocol as 'transitional substances', with a phase-out date of 2030. HFCs contribute much less than CFCs to global warming and have no ozone depletion potential because they contain neither chlorine nor bromine.

In order to investigate the potential environmental impacts of fluorocarbon alternatives to CFCs, a co-operative research programme, the Alternative Fluorocarbons Environmental Acceptability Study (AFEAS), was set up by the major producers. The principal HCFC and HFC alternatives targeted for investigation by AFEAS in co-operation with government research programmes are listed in Table 1.

The conclusion of the initial AFEAS assessment, published in 1989[12] by the World Meteorological Organization (WMO) as an appendix to their scientific assessment of stratospheric ozone, was that the proposed substitutes would be significantly better than the current CFCs, with much smaller potentials for

---

[10] M. McFarland and J. Kaye, *Photochem. Photobiol.*, 1992, **55**, 911.

[11] DuPont estimate, quoted in reference 10 and by F. A. Vogelsberg, Jr, 'CFC Phaseout & Global Climate Change: Challenges for Air Conditioning and Refrigeration', US Congressional Staff Briefing, June 13, 1994.

[12] World Meteorological Organization, 'Scientific Assessment of Stratospheric Ozone: 1989', Global Ozone Research and Monitoring Project—Report no. 20, volume II, 'Appendix: AFEAS Report', WMO, Geneva, 1989.

Table 1 Principal HCFC and HFC alternatives to CFCs

HCFCs		HFCs	
22	$CHClF_2$	32	CH_2F_2
123	$CHCl_2CF_3$	125	CHF_2CF_3
124	$CHClFCF_3$	134a	CH_2FCF_3
141b	CH_3CCl_2F	143a	CH_3CF_3
142b	CH_3CClF_2	152a	CH_3CHF_2
225ca	$CHCl_2CF_2CF_3$		
225cb	$CHClFCF_2CClF_2$		

ozone depletion and greenhouse warming. Although the breakdown products would include acidic compounds, the contribution of these to acid deposition would be insignificant. Moreover, the HCFCs and HFCs would not contribute to photochemical smog formation in urban areas. These conclusions justified proceeding towards commercialization, whilst at the same time conducting research to ensure environmental acceptability.

Subsequent research has greatly enhanced understanding of the atmospheric behaviour of HCFCs and HFCs, making it possible to base judgements of their environmental acceptability on firm, scientific evidence. Extensive review of the research results[4,5,13-15] has not significantly altered the overall conclusions of the 1989 WMO review, as illustrated in the rest of this chapter.

2 Reaction Chemistry and Atmospheric Lifetimes of HCFCs and HFCs

Atmospheric Degradation Processes

The dominant loss process for HCFCs and HFCs in the atmosphere is by reaction with the hydroxyl radical ($^{\cdot}OH$). Photolysis in the stratosphere contributes to the loss of HCFCs, leading directly to the formation of chlorine atoms, but is not important for HFCs. In general, other loss processes such as reaction with $Cl^{\cdot}$ atoms, $NO_3^{\cdot}$, or $O(^3P)$ atoms are negligible. This is because either the rate coefficients for such reactions are small or the atmospheric abundance of the reactant free radical is small. For some species, reaction with $O(^1D)$ in the stratosphere can contribute a minor, additional loss. Thus, the key inputs for calculating the atmospheric lifetimes of HCFCs and HFCs are the rate coefficients for the reactions of these species with $^{\cdot}OH$ and their UV/visible absorption cross-sections.

[13] 'Proceedings of the STEP-HALOCSIDE/AFEAS Workshop: Kinetics and Mechanisms for the Reactions of Halogenated Organic Compounds in the Troposphere', Dublin, March 1993; ed. H. W. Sidebottom, University College, Dublin, 1993.

[14] 'Proceedings of the STEP-HALOCSIDE/AFEAS Workshop: Kinetics and Mechanisms for the Reactions of Halogenated Organic Compounds in the Troposphere', Dublin, May 1991; ed. H. W. Sidebottom, University College, Dublin, 1991.

[15] 'Proceedings of the NASA/NOAA/AFEAS Workshop: Atmospheric Degradation of HCFCs and HFCs', Boulder, Colorado, November 1993; ed. A. McCulloch, P. M. Midgley, and A.-M. Schmoltner, AFEAS, West Tower—Suite 400, 1333 H Street NW, Washington, DC 20005, USA, 1995.

Table 2 Atmospheric lifetimes, ozone depletion potentials, and halocarbon global warming potentials* of CFCs, HCFCs, and HFCs[5,17]

		UNEP Lifetime (years)	ODP (relative to CFC 11)	HGWP (relative to CFC 11)
CFC 11	CCl_3F	50	1.00	1.0
CFC 12	CCl_2F_2	102	0.82	2.9
CFC 113	CCl_2FCClF_2	85	0.90	1.6
CFC 114	$CClF_2CClF_2$	300	0.85	7.1
CFC 115	$CClF_2CF_3$	1700	0.41	35
	CCl_4	42	1.20	0.35
	CH_3CCl_3	5.4	0.12	0.02
HCFC 22	$CHClF_2$	13.3	0.03	0.36
HCFC 123	$CHCl_2CF_3$	1.4	0.01	0.02
HCFC 124	$CHClFCF_3$	5.9	0.03	0.10
HCFC 141b	CH_3CCl_2F	9.4	0.10	0.14
HCFC 142b	CH_3CClF_2	19.5	0.05	0.44
HCFC 225ca	$CHCl_2CF_2CF_3$	2.5	0.02	0.04
HCFC 225cb	$CHClFCF_2CClF_2$	6.6	0.02	0.12
HFC 32	CH_2F_2	6		0.13
HFC 125	CHF_2CF_3	36		0.74
HFC 134a	CH_2FCF_3	14		0.29
HFC 143a	CH_3CF_3	55		1.1
HFC 152a	CH_3CHF_2	1.5		0.03

*Terms are defined in Section 3 of this chapter.

Calculation of Tropospheric Lifetimes. In addition to the ·OH reaction rates, knowledge of the atmospheric abundance of ·OH is also needed. The concentration of ·OH varies significantly in space and time but a globally averaged value, which is applicable for molecules that live longer than a year or so, can be derived from methyl chloroform data. Thus, the average lifetimes of HCFCs and HFCs are calculated by scaling to the average tropospheric lifetime of methyl chloroform, which is estimated independently.[16] The accumulated uncertainties in the atmospheric lifetimes are at least 25–40%. Relative to one another, however, the lifetimes are far less uncertain and are probably known well enough for those compounds currently being produced in environmentally significant quantities. The reference lifetimes adopted by UNEP[5] and by the WMO–UNEP Intergovernmental Panel on Climate Change (IPCC)[17] are shown in Table 2, along with the resultant calculated ozone depletion potentials and global warming potentials (ODPs and GWPs).

At the time of writing, preliminary reports indicate that the calibration factors for atmospheric methyl chloroform measurements may be revised. If these reports are confirmed, the atmospheric lifetime of methyl chloroform would be

[16] 'Report on Concentrations, Lifetimes, and Trends of CFCs, Halons, and Related Species', ed. J. A. Kaye, S. A. Penkett, and F. M. Ormond, NASA Reference Publication no. 1339, 1994.
[17] 'Radiative Forcing of Climate Change: 1994. Report to IPCC from the Scientific Assessment Working Group (WG-I)', WMO–UNEP Intergovernmental Panel on Climate Change, 1994.

reduced by just over 10%, leading to reductions of about the same order in the atmospheric lifetimes of the HCFCs and HFCs. Their calculated ODPs and GWPs would also reduce slightly but the exact amount cannot yet be quantified. This calibration issue urgently needs clarification.

The rate constants for the $\cdot$OH radical reactions with HCFCs and HFCs have been studied in a number of laboratories by a variety of experimental techniques and have been critically evaluated so that they are now reasonably well known.[18,19] The calculated tropospheric lifetimes of HCFCs and HFCs are inversely proportional to their $\cdot$OH radical reaction rates at ~270 K and range from about 2 to less than 40 years. Although these are all much shorter than for the CFCs, the reactivities of HCFCs and HFCs are too low for them to contribute significantly to tropospheric ozone formation.[12]

Reaction with $\cdot$OH. The reaction of $\cdot$OH radicals with HCFCs and HFCs proceeds by hydrogen atom abstraction to produce haloalkyl radicals. For example, taking the generalized structure CX_3CYZH, where X, Y, and Z are H, Cl, Br, and/or F:

$$\cdot OH + CX_3CYZH \rightarrow H_2O + CX_3\dot{C}YZ \tag{1}$$

Under atmospheric conditions, the haloalkyl radicals form the corresponding peroxyl radicals ($CX_3CYZOO\cdot$) and the subsequent reactions are shown in Scheme 1. If the resultant carbonyl $CX_3C(O)Y$ is an aldehyde (*i.e.* Y = H), a further reaction sequence can occur as shown in Scheme 2. The $\cdot CX_3$ radical produced in this sequence then forms a peroxyl radical and undergoes a series of reactions analogous to those outlined for the $CX_3\dot{C}YZ$ radical in Scheme 1.

Peroxyl radicals can react with NO, NO_2, or $HO_2\cdot$ radicals under tropospheric conditions. The available evidence suggests that the major loss process is by reaction with NO to give the alkoxyl radical, $CX_3CYZO\cdot$:

$$CX_3CYZOO\cdot + NO \rightarrow CX_3CYZO\cdot + NO_2 \tag{2}$$

There is no evidence to date of nitrate formation via an alternate pathway to reaction 2. The hydroperoxides, pernitrates, and nitrates formed in the other reactions in Scheme 1 have been shown to have limited stability and their transport to the stratosphere will be negligible. Thermal decomposition or photolysis of these species leads either to regeneration of $CX_3CYZOO\cdot$ or formation of $CX_3CYZO\cdot$.

[18] R. Atkinson, D. L. Baulch, R. A. Cox, R. F. Hampson, Jr, J. A. Kerr, and J. Troe, 'Evaluated Kinetic and Photochemical Data for Atmospheric Chemistry: Supplement IV', *J. Phys. Chem. Ref. Data*, 1992, **21**, 1125.
[19] W. B. DeMore, S. P. Sander, D. M. Golden, R. F. Hampson, M. J. Kurylo, C. J. Howard, A. R. Ravishankara, C. E. Kolb, and M. J. Molina, 'Chemical Kinetics and Photochemical Data for Use in Stratospheric Modeling: Evaluation Number 10', Publication 92-20, Jet Propulsion Laboratory, California Institute of Technology, 1992.

Scheme 1 Atmospheric degradation of HCFCs and HFCs with the generalized structure CX_3CYZH

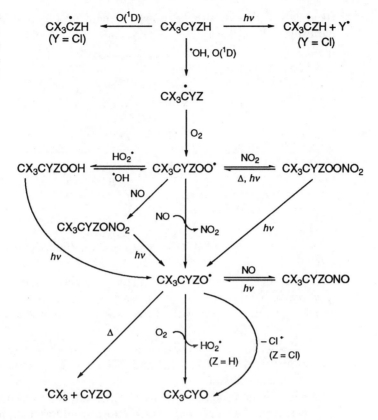

Scheme 2 Atmospheric degradation of haloaldehydes with the generalized structure CX_3CYO

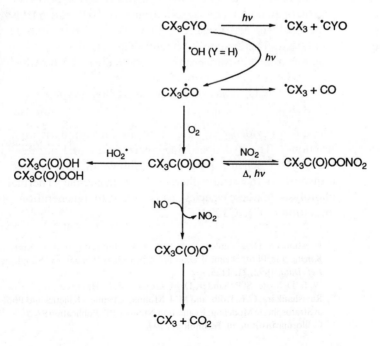

P.M. Midgley

Degradation of Haloalkoxyl Radicals. There are a number of possible reaction pathways:

C–Cl bond cleavage

$$CX_3CYClO^{\bullet} \rightarrow CX_3CYO + Cl^{\bullet} \qquad (3)$$

C–C bond cleavage

$$CX_3CYZO^{\bullet} \rightarrow {}^{\bullet}CX_3 + CYZO \qquad (4)$$

Hydrogen atom abstraction

$$CX_3CYHO^{\bullet} + O_2 \rightarrow CX_3CYO + HO_2^{\bullet} \qquad (5)$$

Review of the now considerable body of experimental data concerning the reactions of haloalkoxyl radicals gives rise to a number of general conclusions about the relative importance of the various reaction channels:[13,20]

1. $CX_2ClO^{\bullet}$ and $CX_2BrO^{\bullet}$ radicals (X = H, Br, Cl, or F) eliminate a $Cl^{\bullet}$ or $Br^{\bullet}$ atom, except for $CH_2ClO^{\bullet}$ and $CH_2BrO^{\bullet}$ where reaction with O_2 is the dominant reaction.
2. $CH_2FO^{\bullet}$ and $CHF_2O^{\bullet}$ radicals react with O_2 to give the corresponding carbonyl fluorides and $HO_2^{\bullet}$ radicals. Fluorine atom elimination or reaction with O_2 are unimportant for $CF_3O^{\bullet}$ radicals. Loss of $CF_3O^{\bullet}$ is largely determined by reaction with CH_4 and nitrogen oxides.
3. $CX_3CH_2O^{\bullet}$ radicals (X = H, Cl, or F) react predominantly with O_2 to give the aldehyde and $HO_2^{\bullet}$ radicals.
4. $CX_3CCl_2O^{\bullet}$ and $CX_3CFClO^{\bullet}$ radicals decompose by $Cl^{\bullet}$ atom elimination rather than C–C bond fission.
5. $CX_3CF_2O^{\bullet}$ radicals undergo C–C bond fission.
6. $CX_3CHYO^{\bullet}$ radicals (Y = Cl or F) have two important reaction channels. The relative importance of C–C bond fission and reaction with O_2 is a function of temperature, O_2 pressure, and total pressure, and hence varies considerably with altitude.

Primary Degradation Products

The principal products expected from HCFCs and HFCs are thus:

acid halides	$CF_3C(O)Cl$ and $CF_3C(O)F$
carbonyl halides	CF_2O, $CFClO$, CCl_2O, HCFO, and HCClO
aldehydes	CX_3CHO

[20] H. W. Sidebottom, in reference 15, p. 2-9.

Table 3 Principal products of atmospheric degradation of HCFCs and HFCs

Parent Compound		Products
HCFC 22	$CHClF_2$	HCl, COF_2 (HF, CO_2)
HCFC 123	$CHCl_2CF_3$	HCl, $CF_3C(O)OH$
HCFC 124	$CHClFCF_3$	HF, HCl, $CF_3C(O)OH$
HCFC 141b	CH_3CCl_2F	CCl_2FCHO (HF, HCl, CO_2)
HCFC 142b	CH_3CClF_2	$CClF_2CHO$ (HF, HCl, CO_2)
HCFC 225ca	$CHCl_2CF_2CF_3$	HCl, $CF_3CF_2C(O)OH$
HCFC 225cb	$CHClFCF_2CClF_2$	HF, HCl, $CClF_2CF_2C(O)OH$
HFC 32	CH_2F_2	COF_2 (HF, CO_2)
HFC 125	CHF_2CF_3	COF_2 (HF, CO_2)
HFC 134a	CH_2FCF_3	COF_2, HCOF (HF, CO_2), $CF_3C(O)OH$
HFC 143a	CH_3CF_3	CF_3CHO
HFC 152a	CH_3CHF_2	COF_2 (HF, CO_2)

For some of these products [$CF_3C(O)Cl$, HCClO, and CX_3CHO], photolysis in the troposphere will be an important removal process. For the other species listed above, physical removal from the atmosphere will be crucial in preventing their accumulation there. The chemical reactions by which HCFCs and HFCs are oxidized in the atmosphere are now sufficiently well-known that it is possible to assign degradation mechanisms and final products, as shown in Table 3, with confidence. As well as stable halogenated products, HCl, HF, and CO_2 also result from decomposition of HCFCs and HFCs. However, the amounts produced will be insignificant compared to natural fluxes.

Removal Processes for Stable Intermediates

A wide variety of atmospheric processes could participate in the removal of the relatively stable, partially oxidized degradation products of HCFCs and HFCs including: uptake by both tropospheric and stratospheric aerosols; uptake by cloud droplets; rainout (wet deposition); and dry deposition to vegetation, to soil and dew surfaces, and to the oceans. However, recent model analyses indicate that uptake by cloud droplets and the oceans are the most likely rate-determining removal processes.[21] The physico-chemical processes which are important in determining the rate of atmospheric trace gas uptake by liquid surfaces include gas phase diffusion, mass accommodation, solvation (Henry's law solubility), liquid phase diffusion, and liquid phase reaction. For highly soluble or reactive species, uptake is likely to be controlled by gas phase transport or mass accommodation, while the uptake of less labile species will be controlled by solubility and/or liquid phase reaction.

Halogenated Aldehydes and Acyl Peroxynitrates. Aldehydes of the generalized structure CX_3CHO are formed as primary products in the ·OH-initiated oxidation of HCFCs and HFCs with the structure CX_3CH_3. The aldehydes further degrade as shown in Scheme 2 (page 97) by photolysis, forming ·CX_3

---

[21] C.E. Kolb, D.R. Worsnop, and O. Rattigan, in reference 15, p. 4-1.

radicals, or by reaction with 'OH radicals, with typical lifetimes of the order of several days. For these compounds, photolysis rather than any heterogeneous process will be the dominant removal process.[13]

The minor channel involving reaction with 'OH radicals can lead to formation of acyl peroxynitrates which appear to be thermally stable under conditions typical of the upper troposphere. Their lifetimes will be controlled by photolysis and can be approximated to that of $CH_3C(O)OONO_2$. The yields of the acyl peroxynitrates will be largely dependent on the stability of the precursor acyl radicals, $CX_3\dot{C}O$. Kinetic studies have shown that, whereas $CF_3\dot{C}O$ is relatively stable and leads to the formation of $CF_3C(O)OO\cdot$, $CCl_3\dot{C}O$ rapidly decomposes. The possibility that $CF_2Cl\dot{C}O$ and $CFCl_2\dot{C}O$ radicals (from the oxidation of CF_2ClCH_3 and $CFCl_2CH_3$) could be sufficiently stable to produce small amounts of the acyl peroxynitrates $CF_2ClC(O)OONO_2$ and $CFCl_2C(O)OONO_2$ was examined because these products, along with CF_3Cl, are potential carriers of chlorine into the stratosphere. There are still uncertainties about the absolute magnitudes of the yields of these compounds because of uncertainties about the quantum yield of photolysis of the haloaldehydes and about the stability of $CX_3\dot{C}O$. However, it is now clear that these yields are in fact very small[15] and thus these minor degradation products will not materially affect the ODPs or GWPs of the parent compounds.

Carbonyl Halides. Reactions of 'OH radicals with the fully halogenated carbonyl compounds CCl_2O, $CClFO$, CF_2O, CX_3CFO, and CX_3CClO have been shown to be unimportant and absorption cross-section measurements indicate that photolysis in the troposphere is only important for CCl_2O and CX_3CClO. Both photolysis and reaction with 'OH radicals may be important for $CHClO$ in the troposphere, but both these sinks are negligible for $CHFO$.

In the troposphere, the carbonyl halides and acid halides are likely to be removed by heterogeneous processes. Uptake by cloud droplets is a critical removal process. The lifetime towards this uptake is inversely proportional to the product of the Henry's constant (H) and the hydrolysis rate constant (k_{hyd}):[21]

$$\tau_{uptake} \propto [H.k_{hyd}]^{-1} \qquad (6)$$

These constants having now been determined for most of the compounds, the estimated lifetimes suggest that the major sink is by uptake in atmospheric and ocean waters. Laboratory measurements of $H.k_{hyd}$ result in calculated lifetimes towards uptake shorter than 15–30 days and can be used to define the geographically averaged deposition. Regional deposition studies, however, require better precision; measurements of $H.k_{hyd}^{\frac{1}{2}}$, H, or k_{hyd} for CF_2O, CF_3CClO, and CF_3CFO disagree by factors of between 5 and 15. There is also disagreement about the extent of acid catalysis of the hydrolysis. Although these disagreements will not be important for the calculated global concentrations of the decomposition products, they have the potential to affect local concentrations.

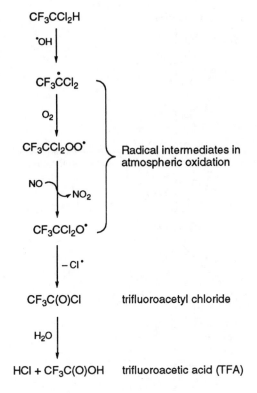

Scheme 3 Atmospheric decomposition of HCFC 123

CF_3CCl_2H

$^\cdot OH$

$CF_3\overset{\cdot}{C}Cl_2$

O_2

$CF_3CCl_2OO^\cdot$ — Radical intermediates in atmospheric oxidation

NO / NO_2

$CF_3CCl_2O^\cdot$

$- Cl^\cdot$

$CF_3C(O)Cl$ — trifluoroacetyl chloride

H_2O

$HCl + CF_3C(O)OH$ — trifluoroacetic acid (TFA)

Trifluoroacetic Acid (TFA)

TFA Formation. Trifluoroacetic acid [$CF_3C(O)OH$ or TFA], the hydrolysis product of both CF_3CFO and CF_3CClO, has been identified as a probable product of the atmospheric degradation of a number of alternative fluorocarbons. CF_3CClO is derived from HCFC 123 and CF_3CFO is formed in the breakdown of HCFC 124 and HFC 134a. It has been calculated that the yield of CF_3CFO will be approximately 40% from HFC 134a and 100% from HCFC 124.[15] Both trifluoroacetyl halides react with water droplets in clouds or surface waters, forming TFA and hydrofluoric or hydrochloric acids, as shown in the simplified reaction schemes for the atmospheric degradation of HCFC 123 (Scheme 3) and HFC 134a (Scheme 4).

As mentioned in the discussion of carbonyl halides above, trifluoroacetyl chloride can be photolysed in the troposphere. This loss process is not open to trifluoroacetyl fluoride and TFA, although they are photolysed at stratospheric wavelengths. The principal process to purge these compounds from the atmosphere involves physical removal in water. When it rains, the TFA in clouds will rain out in the form of the trifluoroacetate ion and in surface waters it will be present as TFA.

The uptake of TFA and other haloacetic acids is not limited by the Henry's constant.[21] Measurements of sticking coefficients indicate that uptake will be as rapid as contact with atmospheric liquid water. It is not known what happens to TFA or other haloacetic acids when cloud water droplets evaporate. Some

101

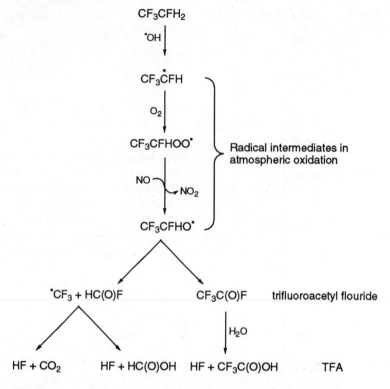

Scheme 4 Atmospheric decomposition of HFC 134a

experiments suggest vaporization or destruction by reaction as the droplets become progressively more acid, but these processes may not occur in real atmospheric droplets where the concentration of counter-ions may be much higher than that of TFA.

There is no known natural source of TFA and, at the present time, its environmental fate is uncertain. Trifluoroacetate appears to be chemically very stable under environmental conditions, generally reacting, if at all, much more slowly than the other haloacetates.[22]

Biological Studies of TFA. TFA has little or no toxicity to animals in either acute or chronic exposures. Although it had been reported to be mildly herbicidal, the literature was incomplete and somewhat questionable. In order to provide relevant information at the concentrations, typically $\mu g\,l^{-1}$, that might be expected in the environment, AFEAS established a programme of research to examine the potential interactions of TFA with the biological environment.[23]

Tests have been conducted on fish, crustacea, and a wide range of plants and algae. In all but one case, the no-effect concentrations were many hundreds of times greater than the anticipated environmental concentrations. The exception, an alga *selenastrum capricornutum* which temporarily and reversibly stopped

[22] S. Madronich and F. J. Dentener, in reference 15, p. 5-1.
[23] Alternative Fluorocarbons Environmental Acceptability Study (AFEAS), 'Proceedings of the Workshop on the Environmental Fate of Trifluoroacetic Acid', Miami Beach, Florida, 3–4 March 1994; AFEAS, West Tower—Suite 400, 1333 H Street NW, Washington, DC 20005, USA, 1994.

growing when subjected to TFA at a concentration of $0.12 \, mg \, l^{-1}$ (between 10 and 100 times greater than the maximum anticipated concentration)[24] is the subject of on-going, further study.

Although high concentrations of TFA have seemed resistant to biodegradation, a recent study has indicated that TFA may be degraded in methanogenic systems at environmentally relevant concentrations.[25] There are conflicting reports on the extent of degradation of environmental TFA by microbial ecosystems in sediments. Further work in this area is needed.

Delivery Dosage of TFA.   While the globally averaged deposition of trifluoroacetate may be estimated simply from a mass balance on the quantities of precursors which may be released, pronounced regional variations are anticipated. Calculations based on kinetic data, published scenarios for future emissions of the precursor HCFCs and HFCs, and atmospheric two-dimensional (2D) models have been carried out to estimate future TFA deposition rates. The principal findings of one such calculation[26] were that, for the emission scenario assumed by the US EPA of 221 000 metric tonnes per year of HFC 134a, 50 000 metric tonnes per year of HCFC 124, and 25 000 metric tonnes per year of HCFC 123 in the year 2010, the globally averaged concentration of TFA in precipitation would be less than about $0.2 \, \mu g \, l^{-1}$. In this calculation, some 15% of the TFA would arise from HCFC 123 releases, 20% from HCFC 124, and 65% from HFC 134a. Due to variations in temperature and rainfall, TFA concentrations will show considerable latitudinal, longitudinal, and seasonal variations. For a given latitude and season, the TFA concentration will also vary inversely in proportion to the amount of rainfall.

Trifluoroacetate has been observed in air samples from southern Germany[27] at concentrations higher than could be explained by current releases of the HCFC and HFC precursors. There is, as yet, no convincing explanation for the presence of TFA in the atmosphere there. Further work is needed to establish the validity of these observations and to determine whether TFA is present only in that area or is distributed more widely.

3   Environmental Effects of HCFCs and HFCs

Effects on the Troposphere

Expected Tropospheric Concentrations.   Using published emission scenarios for likely future production of HCFCs and HFCs, it has been calculated that the tropospheric concentration of most HCFCs and HFCs will remain below 100 ppt;[28] HCFC 22 will be somewhat higher. Based on the recent kinetic information reviewed in the preceding section, it is also calculated that none of the relatively stable intermediate oxidation products such as carbonyl compounds and acid halides will build up substantially in the atmosphere. Local concentrations

[24] R. S. Thompson, in reference 23.

[25] P. T. Visscher, C. W. Culbertson, and R. S. Oremland, *Nature (London)*, 1994, **369**, 729; and also in reference 23.

[26] T. K. Tromp, J. M. Rodriguez, M. K. W. Ko, C. W. Heisey, and N. D. Sze, in reference 23.

[27] H. G. C. Frank, in reference 23.

[28] M. Kanakidou, F. J. Dentener, and P. J. Crutzen, submitted to *J. Geophys. Res.*, 1994.

of TFA, the stable product of the atmospheric degradation of certain fluorocarbon alternatives, may be in the $\mu g l^{-1}$ range in precipitation.[26]

Photochemical Smog. In the 1989 AFEAS report, it was shown that HCFCs and HFCs would not contribute to photochemical smog formation in urban areas.[29] Based on the emission rates projected then, higher than would now be projected, it was calculated that the contribution of all the HCFCs and HFCs together would be less than 0.01% of the contribution of natural ozone precursors.

Acid Deposition. The quantities of chloride and fluoride produced in the atmospheric degradation of HCFCs and HFCs are trivial compared to natural fluxes. For example, assuming emissions of 100 000 tonnes per year of each HCFC (likely to be an extreme over-estimate), it has been calculated that the resulting hydrogen chloride would amount to less than 0.003% of the natural flux. Although the breakdown products of HCFCs and HFCs include acidic compounds, the contribution of these to acid deposition has been calculated to be insignificant.[12]

Ozone Depletion Potentials (ODPs)

Ozone depletion (and halocarbon global warming) potentials are calculated relative to CFC 11. The ODP of a gas is defined as the change in total ozone per unit mass emission of the gas relative to the change in total ozone per unit mass emission of CFC 11. ODPs are currently determined by two different means: by calculations with 2D models of the atmosphere and by a semi-empirical approach.[30] The most recent versions of the 2D models now include representations of heterogeneous chemistry of both the mid-latitude and polar stratosphere. However their treatment of the effects of polar processing are heavily parameterized. The semi-empirical approach, which uses measurements of model compounds in the real stratosphere, provides a useful check on the values derived from 2D theoretical models, and both approaches give similar results.[31]

Values for ODPs, quoted relative to CFC 11, have uncertainties estimated to be $\pm 60\%$ but, within the family of HCFCs, the compound to compound uncertainty is likely to be much less than this. As can be seen from Table 2 (page 95), the reference values of ODPs adopted by UNEP for HCFCs range from 0.01 to 0.10. HFCs, containing neither chlorine nor bromine, do not have ODPs. Although steady-state values have primarily been used in regulatory considerations, ODPs can be defined time-dependently. The original concept of a steady-state ODP was intended to represent the total ozone impact (relative to CFC 11) of a chemical from the time it is emitted to the atmosphere to the time it is completely removed. The more recently proposed concept of a time-dependent ODP (TODP) provides information on the near term effects of a short lived species relative to the near term effects of CFC 11.

Compounds such as HCFCs, simply because of their shorter lifetimes, will

[29] H. Niki, in reference 12, p. 409.
[30] S. Solomon, M. J. Mills, L. E. Heidt, and A. F. Tuck, *J. Geophys. Res.*, 1992, **97**, 825.
[31] D. J. Wuebbles, in reference 15, p. 3-1.

release their chlorine much earlier than CFC 11 does. The TODPs which can then be calculated for the HCFCs are much larger than their steady-state values. However an unambiguous definition of the calculation is needed in order to distinguish the effects of the reference compound from those of the study compounds. For short-lived species, the TODP tends to decrease rapidly with time. A large value of TODP during the first few years does not necessarily imply a significant overall impact on ozone because the TODP relates to the first few years after initial release, when the ozone impact due to CFC 11 itself is exceedingly small. The short term impact of a compound will depend on the actual amount of the material emitted over that time period.

Emission estimates, as developed by industrial producers of halocarbons,[32] provide valuable checks on modelled lifetimes and hence on understanding of atmospheric processes. This understanding will improve as data become available on a broader range of compounds. Emission estimates for the newer HCFCs are needed for comparison with atmospheric measurements as they become available.

Stratospheric Effects of Intermediates

It is now clear that fluorinated radical intermediates in the decomposition process of HCFCs or HFCs, such as $FO^{\cdot}$, $F\dot{C}O$, or $^{\cdot}CF_3$, will not alter the ODP (or GWP) of the source compounds. Several catalytic cycles involving fluorinated free radicals in stratospheric ozone depletion had been postulated but there is now a good body of kinetic data for the key reactions[15] which shows that:

1. Catalytic reactions of $CF_3O_x^{\cdot}$ with O_3 are slow.
2. $FO^{\cdot}$ and $FC(O)O^{\cdot}$ have low reactivity towards stratospheric ozone.
3. Reaction of $CF_3O_x^{\cdot}$ with NO to form the sink species CF_2O is fast.
4. The reservoir species CF_3OH is very stable under stratospheric conditions.

Based on these data, the catalytic chain length of the reactions is calculated to be less than unity, implying a negligible influence of fluorine in general, and the $^{\cdot}CF_3$ grouping in particular, on the ODPs of HCFCs and HFCs.

The quantities of chlorinated peroxyacyl nitrates likely to be formed and transported to the stratosphere are trivial compared with the flux of the parent HCFCs into the stratosphere.

Global Warming Potentials (GWPs)

There are many questions about the most appropriate approach for the definition of GWPs.[31] For the purposes of regulation and for life cycle analyses of equipment, activities or processes, the most useful global warming potentials are those referenced to carbon dioxide (CO_2). As defined by the IPCC,[33] the GWP of

[32] P. M. Midgley, *Ber. Bunsen-Ges. Phys. Chem.*, 1992, **96**, 293.
[33] 'Climate Change 1992: The Supplementary Report to the IPCC Scientific Assessment', WMO–UNEP Intergovernmental Panel on Climate Change, Cambridge University Press, Cambridge, UK, 1992.

P.M. Midgley

a gas is the integrated change in radiative forcing per unit mass pulse emission of the gas relative to the same for CO_2. Radiative forcing is defined as the net radiative flux at the tropopause after allowing for the stratosphere to readjust. However there is a fundamental problem with the use of CO_2 as reference gas because current understanding of the carbon cycle is far from adequate. A principal parameter in the calculation of GWPs, the lifetime of CO_2, is both long and uncertain. Uncertainties include:

1. The carbon cycle itself is uncertain and this affects the calculation of CO_2 decay functions directly.
2. The atmosphere is not in equilibrium now, nor will it be in the future, so that the calculated partitioning of emissions of CO_2 between ocean and atmosphere will depend on assumptions about the future composition of the atmosphere.
3. Furthermore, loss processes which are affected by concentration have lifetimes which will vary in response to concentration changes. This applies to methane (CH_4) and nitrous oxide (or dinitrogen oxide, N_2O) as well as to CO_2.

The consequence is that a variety of parameters will have to be calculated and applied to the individual uses for which they are most appropriate. One suggestion is to adopt an alternative definition for GWPs using a different reference gas. A further consideration is the use of a more realistic background scenario, rather than assuming constant background concentrations of CO_2 and other gases. There is currently much debate about the definition of GWPs that will best meet the needs of policy makers.

Halocarbon Global Warming Potentials (HGWPs), *i.e.* GWPs set relative to CFC 11, are shown in Table 2 (page 95). The calculated values, derived from the IPCC radiative forcings,[17] range from 0.02 to 0.44 for the HCFCs and from 0.03 to 1.1 for the HFCs considered. However, these should be compared to the generally higher values for the CFCs which range from 1 for CFC 11, the reference compound, to 35.

In the same way that for ODPs a time-dependent element has been introduced, so for GWPs, consideration of the time scale of emissions has resulted in the concept of an Integration Time Horizon (ITH). Simply put, this involves looking at the impact of a greenhouse gas at a stated time after emission, rather than at steady state. The ITH selected may reflect the life cycle of the equipment, say 15 years. The GWPs of short-lived species such as the HCFCs and HFCs are typically much higher when considered with an ITH of 15–20 years than with an ultimate ITH of 100–500 years. The reverse is true for very long-lived gases such as CFC 115.

For gases other than CO_2, there are still some disagreements about absolute IR absorptions, overlaps between absorption bands, and the effects of clouds on the radiative properties of the atmosphere. Although the effects of these are smaller than the uncertainty due to CO_2, they will affect the values for one HCFC or HFC relative to another.

Stratospheric ozone is itself a greenhouse gas. Thus, by depleting ozone, a CFC

106

or HCFC can indirectly offset its own GWP. In order to calculate the magnitude of these indirect offsets, it would be necessary to assign ozone depletion (resolved latitudinally and vertically) to individual CFCs or HCFCs and then to offset the IR absorbance of ozone changes against that of the halocarbon, with similar resolution. Even using semi-empirical methods, this is a formidable task. Given that the radiative flux near the tropopause can be very sensitive to the chemical profile there, the uncertainty is likely to be very large.

Total Equivalent Warming Impact Analysis

The objective of the Total Equivalent Warming Impact analysis (TEWI), an activity jointly funded by AFEAS and the US Department of Energy,[34] was to take a systems approach to determine the overall global warming contribution of alternatives available to replace CFCs in their major applications during the 1990s. This overall contribution, or TEWI, is the sum of the direct and indirect emissions of greenhouse gases from the application. The direct effect is the contribution of emissions of the gas itself as a greenhouse gas and the indirect effect is the contribution of the CO_2 emissions resulting from the energy required to operate the system over its normal life. For instance, in cleaning applications, direct emissions are solvent losses multiplied by the GWP of the solvent to express them in CO_2 equivalents and indirect emissions are the actual CO_2 emissions from the fuel used in all aspects of the process.

This analysis has shown that replacing CFCs is a major step in reducing TEWI. The relative importance of direct and indirect effects varies among applications, with the contribution of the direct effect ranging from 2 to 98%. In most energy-intensive applications (refrigeration, air conditioning, insulation) for which data exist, HCFC and HFC systems contribute less to global warming (*i.e.* less TEWI) than currently available non-fluorocarbon or not-in-kind (NIK) alternatives, because of increased energy efficiency. For solvent cleaning technologies, there is no clear TEWI difference among the HCFC/HFC and NIK options.

4 Conclusions

The phase-out of fully halogenated CFCs mandated under the Montreal Protocol is already underway. Several options are available for their replacement and it is expected that about one-quarter of future demand will be met by HCFCs and HFCs. These have many similar properties to the CFCs, thereby satisfying the needs of consumers, whilst at the same time being more acceptable environmentally.

This greater environmental acceptability is based on the fact that these partially halogenated alternatives have much smaller ODPs (zero in the case of HFCs) and GWPs than the CFCs they replace. A significant amount of research

---

[34] S. K. Fischer, P. J. Hughes, P. D. Fairchild, C. L. Kusik, J. T. Dieckmann, E. M. McMahon, and N. Hobday, 'Energy and Global Warming Impacts of CFC Alternative Technologies', AFEAS and US Department of Energy, December 1991. (See references 15 or 23 for AFEAS address).

conducted over the last 5 years has confirmed the validity of these alternatives as a means to accelerate the phase-out of CFCs.

It is now well established that the principal atmospheric loss process for HCFCs and HFCs is by reaction with ˙OH radicals and subsequent oxidation in the troposphere. The primary degradation products include acid halides, carbonyl halides, and halo-aldehydes. The removal processes for these products are increasingly well known.

The atmospheric fate of the alternatives to CFCs and their degradation products is a subject of continued research. Results have been and will be extensively reviewed under the on-going UNEP process, as CFCs are phased out and the transition to HCFCs and HFCs continues.

Volatile Organic Compounds in Indoor Air

DERRICK R. CRUMP

1 Introduction

Buildings serve a wide range of functions as homes, offices, hospitals, schools, shopping centres, and factories. All aim to provide an acceptable environment for people by meeting their physical, chemical, and biological needs. In developed countries people typically spend over 90% of their time in an indoor environment and the greater part of this is in their home. As exposure to air pollution is a function of both time and concentration then the significance of the indoor environment for the total exposure of a person to a pollutant can be high because of the time periods involved.

The indoor environment is generally considered to be a safe haven from air pollutants and during outdoor pollution episodes health authorities advise citizens to remain indoors, particularly those at special risk due to existing respiratory or cardiovascular disease. This is because for pollutants such as sulfur dioxide and ozone the predominant source of contamination of the indoor air is infiltration through the building envelope from the ambient environment.[1] Once indoors the pollutant concentration is depleted through adsorption on surfaces such as walls and furnishings and dilution by the indoor air. In consequence the indoor/outdoor ratios of concentrations of sulfur dioxide and ozone during periods when the ambient concentrations are moderate to high are in the range of 0.3–0.5 and 0.1–0.3 respectively. For some other pollutants the predominant sources may be located within the building itself and for these outdoor air, in the form of infiltration and ventilation, is essential to prevent the build-up of unacceptable concentrations. These pollutants include combustion gases produced by the burning of fossil fuels, notably nitrogen dioxide and carbon monoxide, respirable particles produced by tobacco smoke and a wide range of organic compounds.

The exposure of workers in the industrial environment is controlled through the application of occupational exposure limits and in the UK this is enacted through the Control of Substances Hazardous to Health (COSHH) regulations. The concentrations of concern in this environment are much higher than those normally found in ambient air, and control measures aim to protect the worker from ill health due to exposure during the working day. In the non-occupational

[1] T. Godish, 'Air Quality', Lewis Publishers, Michigan, 1991.

109

environment there are further concerns because there are different population groups, such as children and people more at risk from pollutants, and exposure times are often longer. Organic compounds are one type of pollutant found in the air of non-industrial buildings which have been the subject of increasing concern since the 1970s because of their potential to cause health effects similar to those reported in studies of Sick Building Syndrome and to contribute to causes of respiratory and other diseases including cancer.

This chapter considers the sources, concentrations, health effects, and methods of control of the wide range of volatile organic compounds (VOCs) found in the air of our homes and other non-industrial indoor environments. Various studies of indoor air quality have identified more than 250 organic compounds at a level exceeding 1 ppb and many hundreds of additional compounds undoubtedly exist at lower levels.[2] These compounds have a wide range of boiling points and have been classed as very volatile (<0 to 50–100 °C), volatile (50–100 to 240–260 °C) and semi-volatile (250–260 to 380–400 °C).[3] Examples in each group of concern in the indoor environment are formaldehyde, mineral or white spirits, and polyaromatic hydrocarbons respectively.

2 Sources of VOCs in Indoor Air

VOCs in the indoor air can originate from the outdoor air and also sources located indoors. By comparing differences in concentrations in the indoor and outdoor environment it is possible to evaluate the significance of the outdoor air for determining the indoor concentration of any particular VOC. Table 1 compares results of measurements of formaldehyde, and several other VOCs measured in the main bedrooms of 174 homes in the Avon district of England with concentrations at 14 locations outdoors.[4] The total volatile organic compound concentration (TVOC) represents a summation of all the individual VOCs determined with a boiling point range of 50–280 °C. Measurements of TVOC, benzene, toluene, and undecane were undertaken by use of a diffusive sampler with a 4 week exposure period and those for formaldehyde with a 3 day exposure period. Samplers were placed in the homes each month for a period of up to 12 months. Concentrations of formaldehyde, undecane, and TVOCs are ten or more times higher indoors than outside and hence indoor sources predominate. For toluene, the ratio is approximately 3 : 1 indicating a significant outdoor source, but again indoor sources are more significant. In the case of benzene, the indoor concentration is 1.3 times the outdoors and the indoor concentration is greatly influenced by that outdoors. For benzene and toluene the outdoor source is predominantly motor vehicle exhausts.

The indoor air can be polluted by a large number of individual VOCs

J. Namiesnik, T. Gorecki, B. Mozdron–Zabiegala, and J. Lukasiak, *Building and Environment*, 1992, **27**, 339.

[3] World Health Organisation, 'Indoor Air Quality: Organic Pollutants', Euro reports and studies no. 111, WHO Regional Publications, Copenhagen, 1989.

[4] V. Brown, S. Coward, D. Crump, M. Gavin, C. Hunter, J. Llewellyn, and G. Raw, 'The ALSPAC Indoor Air Environment Study', Building Research Establishment report (in the press), BRE, Watford, 1995.

Table 1 Concentration of VOCs in indoor (main bedroom) and outdoor air in Avon, England

| VOC | Indoor concentration (μg m^{-3}) | | | | Outdoor concentration (μg m^{-3}) | | | |
	No. of readings	Mean	Min.	Max.	No. of readings	Mean	Min.	Max.
Formaldehyde	1501	25	1	205	126	2	ND	11
Benzene	1504	8	ND	78	125	5	ND	16
Toluene	1504	42	2	1793	125	12	2	254
Undecane	1504	14	ND	797	125	1	ND	3
TVOC	1505	400	21	8392	125	35	4	317

ND = none detected ($<1 \mu$g m^{-3}); TVOC = total volatile organic compounds.

generated from many different sources. The sources can be divided into those with continuous emissions (long term emission, constant source strength), and discontinuous emissions (short term emission, variable source strength). While the magnitude of emissions from continuous sources often depends on temperature, relative humidity, and sometimes air velocity, and varies within a time scale of months, discontinuous emissions are much more time dependent and may change within hours or minutes. The nature of the emission and the variability of indoor spaces and ventilation conditions results in a dynamic behaviour of VOCs in the indoor environment.[5]

The continuous and discontinuous sources can also be separated into three groups according to their origin: (i) outdoor air (air, soil, and water sources); (ii) man and his activities (body odours, energy production, smoking, household activities, and hobby products); and (iii) materials and equipment (building and renovation materials, furnishings, HVAC systems). The most important sources with regard to materials and equipment are adhesives, caulks, floor coverings, floor sealants, furniture, HVAC systems, insulation materials, lacquers, office machines, paints, particleboard, and wall coverings.[6]

Two reviews have detailed the wide range of VOCs measured from a large number of sources in indoor air; Table 2 gives examples of types of sources and classes of chemicals known to be emitted but this is far from comprehensive.[2,7] The complexity of emissions and differences between similar product types is illustrated by one study that reported emissions from 200 materials used indoors and quantified some 50 compounds belonging to 11 chemical classes.[8] A review of the scientific literature published before 1991 identified 24 types of incident where building materials had acted as major sources of VOCs in the air in one or more buildings and had resulted in complaints by occupants.[9] Table 3

[5] B. Seifert, H. Knoppel, R. Lanting, A. Person, P. Siskos, and P. Wolkoff, 'Strategy for Sampling Chemical Substances in Indoor Air', European Concerted Action, Report 6, EUR 12617 EN, Commission of the European Communities, Luxembourg, 1989.

[6] B. Seifert, in 'Chemical, Microbiological, Health, and Comfort Aspects of Indoor Air Quality—State of the Art in SBS', ed. H. Knoppel and P. Wolkoff, Kluwer Academic Publishers, Dordrecht, The Netherlands, 1992, p. 25.

[7] R. Otson and P. Fellin, in 'Gaseous Pollutants: Characterisation and Cycling', ed. J. Nriagu, J. Wiley and Sons, Chichester, UK, 1992, ch. 9, p. 335.

[8] K. Engstrom, L. Nyman, M. Hackola, and H. Saarri, Proceedings of Healthy Buildings '88, Stockholm, 1988, vol. 3, p. 333.

Table 2 Specific indoor sources of VOCs[2,7]

Compound	Source material(s)
p-Dichlorobenzene	Moth crystals, room deodorants
Styrene	Insulation, textiles, disinfectants, plastics, paints
Benzyl chloride	Vinyl tiles
Benzene	Smoking
Tetrachloroethylene	Dry cleaned clothes
Chloroform	Chlorinated water
1,1,1-Trichloroethane	Dry cleaned clothes, aerosol sprays, fabric protectors
Carbon tetrachloride	Industrial strength cleaners
Aromatic hydrocarbons (toluene, xylenes, ethylbenzene, trimethylbenzenes) and aliphatic hydrocarbons	Paints, adhesives, gasoline, combustion products
Terpenes (limonene, α-pinene)	Scented deodorizers, polishes, fabrics, fabric softeners, cigarettes, food, beverages
PAHs	Combustion products (smoking, woodburning, kerosene heaters)
Acrylic acid esters, epichlorohydrin	Monomers may escape from polymers
Alcohols	Aerosols, window-cleaners, paints, paint thinning, cosmetics, and adhesives
Ketones	Lacquers, varnishes, polish removers, adhesives
Ethers	Resin, paints, varnishes, lacquers, dyes, soaps, cosmetics
Esters	Plastics, resins, plasticizers, lacquer solvents, flavours, perfumes

summarizes the sources, main compounds, and concentrations reported. Problems associated with flooring materials predominate. This is probably due to a combination of the large surface area of these products used in a room and their composition, which tends to be manufactured polymeric materials.

Formaldehyde is released into the indoor air from a variety of sources. For example, it is a constituent of tobacco smoke, combustion gases from gas appliances, disinfectants, and an additive in water-based paints. However significant contamination of indoor air is most likely from products manufactured using urea–formaldehyde (U–F) resins. These include particleboard (flooring, panelling, cabinetry, furniture), medium density fibreboard, hardwood plywood panelling and urea–formaldehyde foam insulation. The formaldehyde release can

[9] H. Gustafsson, 'Building Materials Identified as Major Sources for Indoor Air Pollutants', Swedish Council for Building Research, Document D10, Stockholm, 1992.

Table 3 Cases of building materials producing contamination of the indoor air by VOCs[9]

Case	Source	Compounds	Concentrations
1	Adhesive for carpet	Toluene	$30\,mg\,m^{-3}$ initially, declining to $0.1\,mg\,m^{-3}$ over 4 months
2	Laminated cork floor tiles	Phenol	$13–16\,\mu g\,m^{-3}$
3	Vinyl flooring	Alkyl aromatics including dodecylbenzene	$100–200\,\mu g\,m^{-3}$
4	Vinyl flooring	TXIB*	$100–1000\,\mu g\,m^{-3}$
5	Vinyl flooring	Dodecane TXIB*	$40\,\mu g\,m^{-3}$ (dodecane) $50–70\,\mu g\,m^{-3}$ (TXIB)
6	Carpets	4-Phenylcyclohexene	$29–45\,\mu g\,m^{-3}$
7	Carpets	Styrene	0.9 ppm
8	Rubber floor tiles	Styrene	—
9	Sealant	1,2,5-Trithiepane (from oxidation of 2,2-dithioldiethyl thioether)	—
10	Injected damp proof solution	White spirit	$1.5–6\,mg\,m^{-3}$
11	Creosote impregnated timber	Naphthalene Methylnaphthalenes	$20\,\mu g\,m^{-3}$ $10\,\mu g\,m^{-3}$
12	Damp proof membrane	Naphthalene	$0.2\,mg\,m^{-3}$
13	Vinyl floor on damp concrete	2-Ethylhexanol	$1000\,\mu g\,m^{-3}$
14	Vinyl flooring on floor levelling compound	2-Ethylhexanol Ammonia	$10\,\mu g\,m^{-3}$ —
15	Carpets	2-Ethylhexanol Nonanol, heptanol	$34–138\,\mu g\,m^{-3}$ —
16	Vinyl flooring	Ethyl hexyl acrylate	$30\,\mu g\,m^{-3}$
17	Vinyl flooring	Phenol Cresol	$10\,\mu g\,m^{-3}$ $10\,\mu g\,m^{-3}$
18	Paint	Dibutyl phthalate	$40\,\mu g\,m^{-3}$
19	Paint affected by fire	PCB	$1–3\,\mu g\,m^{-3}$
20	Radiator paint	Hexanoic acid Hexanal	$60\,\mu g\,m^{-3}$ $37\,\mu g\,m^{-3}$
21	Paint	White spirit	$1.4\,mg\,m^{-3}$
22	Damp mineral wool insulation	Hexanal	$0.1\,mg\,m^{-3}$
23	Newly built house	130 compounds including styrene TVOC	$0.4\,mg\,m^{-3}$ $13\,mg\,m^{-3}$
24	New house	C_5 and C_6 aldehydes	$50–150\,mg\,m^{-3}$

*TXIB is 2,2,4-trimethyl-1,3-pentadiol diisobutyrate.

be due to free volatilizable unreacted formaldehyde trapped in the resin and from the hydrolytic decomposition of the resin polymer itself.[10]

Environmental tobacco smoke (ETS) is a source of particular concern because of the nuisance and irritation it can cause to users of multi-occupied buildings and the risks of disease to those inadvertently exposed to smoke. It is a complex source consisting of several thousand chemical constituents in gaseous, vapour, and particulate phases. The main VOCs to be released in sidestream smoke in quantities exceeding 1 mg per cigarette are nicotine, acetaldehyde, acrolein, isoprene, and acetonitrile.[11] Very few of the constituents are unique to ETS and this has caused difficulties in apportioning the contribution that ETS makes to concentrations of particular pollutants within buildings.

Studies have shown that where smoking rates are high and ventilation minimal there is a clear contribution to formaldehyde concentrations from ETS of the order of a few 10s of μg m^{-3} and concentrations of aliphatic and monocyclic aromatic hydrocarbons can rise from 2–20 μg m^{-3} to 50–200 μg m^{-3}. Detailed studies in the USA of residential exposures suggest that non-smoking households experience an exposure to approximately 7 μg m^{-3} of benzene, while households with a smoker are exposed to approximately 11 μg m^{-3}. Excess exposure to styrene and xylenes range from 0.5 to 1.5 μg m^{-3}.

3 Measurement of VOCs

The usefulness of any measurement of air quality will depend upon the accuracy of determination of the pollutants collected and the degree to which the sample collected is representative of the concentration in the space or the exposure of an individual under study. In the occupational environment, recommended methods for determining pollutants have been published by regulatory bodies in many countries and a number of national and international standards have been agreed. However, there are few recognized procedures for studying indoor air pollution outside the working environment.

A wide range of sampling and analytical methods have been applied to the measurement of formaldehyde in indoor air.[12] These include types of passive samplers that can be used to provide information on mean concentrations for periods of typically 1–3 days and a variety of active or pumped samplers that provide concentration data for periods of 30 minutes to a few hours.

For the less reactive VOCs the most widely used methods are based on collection using adsorbents contained in a sampling tube or badge.[13] Tenax TA has been the most widely used and this can be thermally desorbed and the VOCs

[10] L. Molhave, 'Indoor Air Pollution by Formaldehyde in European Countries', European Concerted Action, Report 7, EUR 13216 EN, Commission of the European Communities, Luxembourg, 1990.

[11] M. Guerin, R. Jenkins, and B. Thomkins, 'The Chemistry of Environmental Tobacco Smoke', Lewis Publishers, Michigan, 1992.

[12] V. Brown, D. Crump, and M. Gavin, Proceedings of the International Conference on Clean Air at Work, Luxembourg, 1991, p. 357.

[13] H. Knoppel, in 'Chemical, Microbiological, Health, and Comfort Aspects of Indoor Air Quality—State of the Art in SBS', ed. H. Knoppel and P. Wolkoff, Kluwer Academic Publishers, Dordrecht, The Netherlands, 1992, p. 37.

determined by GC with FID, MS, or other detection systems. The sampler can be used in either an active or a diffusive mode. To obtain sufficient sensitivity for analysis the exposure period of the diffusive sampler is typically between 1 and 4 weeks whereas for active sampling this can be achieved with sampling times of 1 hour.

No single adsorbent is suitable for the analysis of the full range of VOCs encountered in indoor air. For example polar compounds and those of molecular weight less than benzene are not readily trapped using Tenax and semi-volatiles boiling about 280 °C are not recoverable at the maximum thermal desorption temperature for Tenax. Other adsorbents include Carbotrap, activated carbon, Porapaks, and organic molecular sieves. Polyurethane foam and XAD resin may be used for collection of semi-volatiles including polyaromatics. Sampling tubes containing 3 layers of different adsorbents have been used in some studies to determine VVOCs, VOCs, and SVOCs in a single analysis, but these are only applicable to active, not diffusive, sampling. Grab or whole air sampling is also used, mostly for analysing VVOCs. Whole air samples are collected in stainless steel canisters and returned to the laboratory for GC analysis.

The important question of where to sample, how often, and for what duration has been addressed in a report published by the European Commission.[14] The recommended sampling strategy depends upon the purpose of the study. For example if the objective is to find out whether indoor air pollution contributes to complaints of acute health effects, sampling should be performed under conditions which enable the measurement of peak concentrations to which an occupant may be exposed. This means that sampling should be performed at a location, a certain point in time, for a certain duration and under environmental conditions yielding maximum exposure levels. On the contrary, the assessment of the potential contribution of VOCs to a chronic effect would require the determination of a time integrated average concentration or total exposure.

4 Concentration of VOCs in Indoor Air

Formaldehyde has been the most widely studied compound in indoor air since reports of health problems in buildings, including mobile homes, in the 1970s were associated with sources such as particleboard and UFFI (urea–formaldehyde foam insulation).[15] In the UK a survey of 178 buildings was undertaken in the early 1980s and found mean concentrations in homes insulated with UFFI of $0.114 \, mg \, m^{-3}$ compared with $0.057 \, mg \, m^{-3}$ in homes without UFFI. Similar concentrations were reported in 90 homes investigated over the period 1985 to 1992.[16] Figure 1 shows the results of monitoring of formaldehyde concentrations in 174 homes in England over a 12 month period related to the age of the building. The formaldehyde concentration was measured once a month using a diffusive

[14] D. Cavallo, D. Crump, H. Knoppel, A. Lauarent, E. Lebret, B. Lundgren, H. Rothweiler, B. Seifert, and P. Wolkoff, 'Sampling Strategies for VOC in Indoor Air', Commission of the European Communities, Luxembourg, 1995.

[15] B. Meyer, 'Indoor Air Quality', Addison–Wesley, Massachusetts, 1983.

[16] V. Brown, A. Cockram, D. Crump, and M. Gavin, Proceedings of Indoor Air '93, Helsinki, vol. 2, 1993, p. 111.

Figure 1 Relationship between age of house and mean annual concentration of formaldehyde in homes in Avon, England

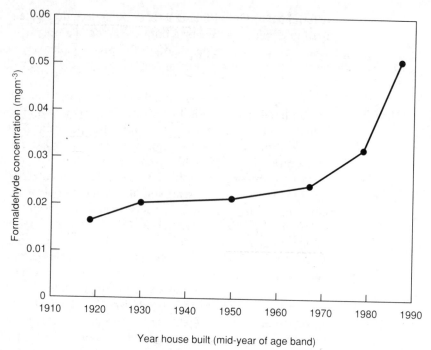

Year house built (mid-year of age band)

sampler and an exposure period of 3 days. The homes were divided into 6 groups based on the age of the house. Homes built in the 1980s had a formaldehyde concentration about three times higher than those built before the 1920s. The higher formaldehyde concentrations are probably due to the relatively high emission from new building and furnishing materials although differences in the amount of ventilation may also be a contributory factor.[4]

Data from other European countries is based mainly on small localized studies or investigations of problem buildings. Average indoor formaldehyde concentrations are in the range of 0.02–0.06 mg m^{-3}.[10] In Australia the 5 day mean concentration was measured on a single occasion in 200 homes; the mean concentration indoors was 0.024 mg m^{-3}. A detailed study of 80 occupants found that measurements in the home accounted for 54–61% of the variation of personal exposure.[17]

The World Health Organization summarized the results of studies of VOC concentrations in homes undertaken in Italy, The Netherlands, the USA, and Germany.[3] Despite differences in sampling times and methods, a single data set was constructed (with an uncertainty of about 50%) that could be considered representative of an average home. Table 4 summarizes the data for the most frequently occurring compounds.

The US Environment Protection Agency investigated routes of exposure of people to VOCs as part of their Total Exposure Assessment Methodology (TEAM) study. Indoor sources and personal activities were shown to outweigh outdoor sources in contributing to personal exposure.[18] Examples of compounds

[17] P. Dingle, S. Hu, and F. Murray, Proceedings of Indoor Air '93, Helsinki, vol. 2, 1993, p. 293.
[18] L. Wallace, *Risk Anal.*, 1993, **13**, 135.

116

Table 4 Percentiles of the frequency distribution of VOC concentrations in the air of private homes (adapted from WHO 1989)[2,3]

Pollutant	Indoor concentration (μg m^{-3})		Outdoor mean
	Median	90-Percentile	
Dichloromethane	< 10	< 10	—
Chloroform	3	15	—
Formaldehyde	25	60	—
Hexanal	1	5	—
Decane	10	50	—
Undecane	5	25	—
Benzene	10	20	3
Toluene	65	150	5
Styrene	1	5	—
Naphthalene	2	5	—
1,1,1-Trichloroethane	5	20	—
Trichloroethylene	5	20	< 2
Tetrachloroethylene	5	20	—
p-Dichlorobenzene	5	20	—
Butanol	< 1	3	—
α-Pinene	10	20	—
Limonene	15	70	—
Ethyl acetate*	5–10	10–50	—
n-Butyl acetate*	2–5	5–10	—

*Values based on Seifert 1992[6].

with strong indoor sources were benzene (smoking), tetrachloroethylene (dry cleaned clothes), chloroform (volatilization from chlorinated water supplies), and trichloroethylene (unknown). The study of 174 homes in the UK found that the mean number of detectable compounds was 85 in indoor air compared with 14 in the outdoor air.[4] Occupant activities, particularly painting and decorating, were shown to have an important effect on the monthly mean VOC concentration indoors.

5 Health Effects of VOCs

Several working groups have evaluated the health effects of formaldehyde.[10] The International Agency for Research on Cancer (IARC) have classified formaldehyde in Group 2B which means there is sufficient evidence of carcinogenicity in animals but inadequate evidence of carcinogenicity in humans. Short-term exposure leads to irritation of the eyes, nose, and throat; and exposure-dependent discomfort, lachrymation, sneezing, coughing, nausea, and dyspnoea (laboured breathing).

The effects on humans of short-term exposure at various formaldehyde concentrations are given in Table 5.[19] The odour detection threshold is approximately 0.1 mg m^{-3} with eye and throat irritation starting at about 0.5 mg m^{-3}. There is a sharp increase in irritation and discomfort between 1 and

[19] World Health Organisation, 'Air Quality Guidelines for Europe', European series no. 23, WHO Regional Publications, Copenhagen, 1987.

Table 5 Effects of
formaldehyde on humans
after short-term
exposure[19]

Effect	Formaldehyde concentration ($mg\,m^{-3}$)	
	Estimated median	Reported range
Odour detection threshold (including repeated exposure)	0.1	0.06–1.2
Eye irritation threshold	0.5	0.01–1.9
Throat irritation threshold	0.6	0.1–3.1
Biting sensation in nose and eye	3.1	2.5–3.7
Tolerable for 30 minutes (lachrymation)	5.6	5–6.2
Strong lachrymation, lasting for 1 hour	17.8	12–25
Danger to life, oedema, inflammation, pneumonia	37.5	37–60
Death	125	60–125

$20\,mg\,m^{-3}$ and at concentrations of $30\,mg\,m^{-3}$ and higher there is a danger to life. Individuals vary widely in the levels of formaldehyde they can tolerate in ambient air.

A number of studies point to formaldehyde as a potential factor predisposing certain groups, particularly children, to respiratory tract infections. It has not been definitely established that formaldehyde gas causes respiratory tract allergy, but occupational studies indicate that 1–2% of the population exposed to high concentrations may develop asthma.[19] A study in Arizona, USA reported that children exposed to 60–120 ppb formaldehyde at home, and particularly if also exposed to tobacco smoke, had an increased likelihood of suffering from asthma or chronic bronchitis.[20]

The health effects of the wide range of other organic compounds that can occur in indoor air at concentrations well below standards used to control occupational exposure can be classed as: (a) odour and other sensory effects such as irritation; (b) mucosal irritation and other morbidity due to systemic toxicity; and (c) genotoxicity and carcinogenicity.[3] The present understanding of these effects is limited by inadequate knowledge of both the population exposure to VOCs and exposure–response relationships.

Many chemical compounds have both odorant and irritant properties. Five main types of sensory systems that respond to irritants are situated on or near the body surface; systems related to the eye, nose, throat, facial skin, and other body skin. Some of these systems tend to respond to an accumulated dose and their reaction is not so immediate and acute as in the case of odour perception. Effects of irritation are numerous and may include conjuctivitis, sneezing, coughing, hoarseness, a feeling of dryness of mucous membranes, skin erythrema or oedema, and changes in breathing patterns. Odour sensation may lead to a number of secondary effects such as vomiting, escape behaviour, triggering of hypersensitivity reactions, and changes in breathing patterns. Exposure studies

[20] M. Kryzanowski, J. Quackenboss, and M. Lebowitz, *Environ. Res.*, 1990, **52**, 117.

using chambers have investigated the physiological effects of mixtures of VOCs and the effects of chemical exposure and air temperature have been shown to interact.[21]

Systemic toxic effects of VOCs include haematological, neurological, hepatic, renal effects, and mucosal irritation. Benzene causes aplastic anaemia and polycythaemia and dichloromethane produces carboxyhaemoglobin. Dichloromethane, toluene, styrene, trichloroethylene, and tetrachloroethylene are neurotoxic. Styrene also produces mucous membrane irritation as does naphthalene.

Genotoxicity and carcinogenicity are effects that inherently express themselves a long time after exposure to a toxic substance. It is assumed that there is no threshold concentration for effect, and risk estimation is therefore performed down to very low concentrations. Five compounds commonly found in indoor air possess particular genotoxic and/or carcinogenic properties: benzene, tetrachloromethane, chloroform, 1,2-dichloroethane, and trichloroethylene (trichloroethene).

6   Control of Indoor Air Pollution due to VOCs

In the industrial environment pollution of the indoor air is controlled through regulations that define maximum exposure limits and occupational exposure standards for many individual VOCs. For example in the UK the maximum exposure limit for formaldehyde is 2 ppm ($2.5\,\mathrm{mg\,m^{-3}}$) for 10 minute and 8 hour exposure periods. There is also an occupational standard for white spirit which is a mixture of VOCs commonly used as solvents for paints, varnishes, and other products.[22] No standards exist for non-occupational environments although a number of authors and groups have considered the usefulness and practical difficulties of establishing either indoor air quality standards or guidelines.[23-25] The WHO has recommended air quality guidelines for outdoor and indoor air based on an assessment of health effects of 28 pollutants including sensory effects.[18] For formaldehyde they recommend that 'in order to avoid complaints of sensitive people about indoor air in non-industrial buildings, the formaldehyde concentration should be below $0.1\,\mathrm{mg\,m^{-3}}$ as a 30 minute average, and this is recommended as an air quality guideline value'.

For human carcinogens the WHO does not recommend acceptable concentrations, but rather the unit risk (cancer risk estimate for lifetime exposure to a concentration of $1\,\mu\mathrm{g\,m^{-3}}$). The guidelines for individual compounds do not take account of possible synergistic effects which are likely to be especially important in the case of mixtures of VOCs.

Guidelines for TVOC concentrations in indoor air have been proposed:

21   L. Molhave, Z. Liu, A. Jorgensen, O. Pedersen, and S. Kjaergaard, Proceedings of Indoor Air '93, Helsinki, vol. 3, 1993, p. 155.
22   Health and Safety Executive, 'Occupational Exposure Limits 1994', EH 40/94, HSE, London.
23   A. Moghissi, *Environ. Int.*, 1991, **17**, 365.
24   B. Seifert, in 'Chemical, Microbiological, Health, and Comfort Aspects of Indoor Air Quality—State of the Art in SBS', ed. H. Knoppel and P. Wolkoff, Kluwer Academic Publishers, Dordrecht, The Netherlands, 1992, p. 311.
25   House of Commons, 'Environment Committee Sixth Report, Indoor Pollution', HMSO, UK, 1991.

$200\,\mu g\,m^{-3}$ for a comfort range and a target guideline of $300\,\mu g\,m^{-3}$.[26] However the analytical methodology used to define the TVOC value is different for the two suggested guidelines. In Australia, the National Health and Medical Research Council has recommended $500\,\mu g\,m^{-3}$ as the level of concern for TVOCs with no single compound contributing to more than 50% of the total.[27]

The provision of good air quality in non-industrial buildings has been addressed in developed countries through requirements for adequate air supply and means of ventilation, as well as standards to ensure efficient working of flues or chimneys for heat producing appliances. In the UK this is achieved primarily through the Building Regulations (parts F and J) which also cover precautions against substances released from the ground (part C) and the ingress of toxic fumes (part D), in particular formaldehyde from urea–formaldehyde foam insulation for cavity walls.[28] Other controls are via prohibition of use of substances (e.g. asbestos) and restrictions on use (e.g. PCPs for remedial treatment of timber products). There are also product standards such as those relating to the free formaldehyde content of particleboard and fibreboard and the quality and installation procedure for use of UFFI for cavity walls. These standards limit the amount of formaldehyde emitted from the product into the occupied space of a building.

To be effective the air used for ventilation must be of good quality. The primary purpose of ventilation is to satisfy the metabolic requirements of occupants and dilute and disperse occupant generated pollutants, though it can also be used to control non-occupant generated pollutants.[29] EC directives set limits on maximum concentrations of some pollutants in ambient air though these do not yet extend to VOCs. In the UK an expert panel has recommended an air quality standard for one VOC, benzene. This is for an annual average maximum concentration of 5 ppb ($16\,\mu g\,m^{-3}$) with a recommendation that this should be reduced in the future to 1 ppb.[30]

The process of ventilation can contribute significantly to the energy demand of a building. This is a significant factor on a global scale because of the role of energy consumption in the generation of atmospheric pollutants. For example, energy used in the housing sector in Western European countries accounts for 28% of total energy demand and space heating accounts for 60% of the energy used in homes.[31] There is therefore a conflict between the need for high ventilation rates to provide good air quality and lower rates to minimize heat loss and prevent occupant discomfort. In mechanically ventilated buildings, particularly in warm climates, energy may also be used for ventilation to reduce heat loads.

[26] D. Bienfait, P. Fanger, K. Fitzner, M. Jantunen, T. Lindvall, E. Skaret, O. Seppanen, J. Schlatter, and M. Woolliscraft, 'Ventilation Requirements in Buildings', European Concerted Action, Report 11, EUR 14449 EN, Commission of the European Communities, 1992.

[27] P. Dingle and F. Murray, Indoor Environ., 1993, 2, 217.

[28] D. Crump, Proceedings of the Investigation of Air Pollution Standing Conference, paper 15/3, London, December, 1993.

[29] M. Limb, 'Ventilation and Building Airtightness', Technical Note AIVC 43, Air Infiltration and Ventilation Centre, Coventry, 1994.

[30] Department of the Environment, 'Benzene', Expert Panel on Air Quality Standards, HMSO, London, 1994.

[31] G. Henderson, Proceedings of Innovative Housing '93, Vancouver, vol. 1, 1993, p. 3.

The Scandinavian countries, Canada, and Switzerland have standards that require lower ventilation rates than other European countries such as the UK which have milder climates.[29] It is probably no coincidence that these countries have reported particular problems due to pollution by organic compounds in indoor air and have been among those instrumental in developing controls on products to limit the emission of pollutants into the indoor air.

A report published by the European Commission notes that compliance with existing ventilation guidelines and standards has not prevented exposure to pollutants with potentially adverse effects and widespread complaints about indoor air quality in many buildings.[26] Materials in a building are often more important polluters than the occupants and contribute significantly to complaints. The report offers a guideline for determining the required ventilation rate in the building based on the total pollution load caused by materials in the building, occupants, and activities. A number of particular problems limit the application of this approach at this time: knowledge of the pollutant emission characteristics of products, the extrapolation of data about emissions to concentration in a building, and an understanding of the dose–response relationship for the pollutants measured in indoor air.

7 Emission Testing of Products

Environmental test chambers are being applied by a growing number of laboratories to characterize the type and amounts of VOCs released by building and consumer products. Guidelines exist for the testing of formaldehyde release from wood based products using room size chambers[32] and for using small chambers ($\leq 1\,m^3$) to investigate the emission of VOCs from a wide range of other products.[33] The principle of such tests is to enclose a sample of the product in an inert enclosure supplied with clean air and maintained at constant temperature and humidity. The air exchange rate through the chamber and the surface area of product to volume of enclosure are selected to be representative of conditions occurring in buildings. VOC concentrations measured in the chamber are used to calculate an emission rate per unit area or weight of the product.

Figure 2 is an example of the GC chromatogram given by analysis of VOCs collected on a Tenax adsorbent tube used to sample air from a $1\,m^3$ chamber containing a sample of vinyl flooring material. This shows the complexity of the emissions involving over 70 detectable peaks with phenol being dominant. The conditions of test were 23 °C and 45% RH (relative humidity) with an air exchange rate of one, a loading ratio of 0.4, and an air speed or 0.3 m s^{-1}. Figure 3 shows how the concentration in the chamber and hence rate of emission change

[32] D. Crump, M. De Bortoli, W. Haelvoet, H. Knoppel, R. Marutzky, P. Nielsen, M. Romeis, and J. Van Der Wal, 'Formaldehyde Emission from Wood Based Materials', European Concerted Action, Report 2, EUR 12196 EN, Commission of the European Communities, Luxembourg, 1989.

[33] A. Colombo, D. Crump, M. De Bortoli, R. Gehrig, H. Gustafsson, P. Nielsen, K. Saarela, H. Sageot, N. Tsalkani, D. Ullrich, and J. Van Der Wal, 'Guideline for the Characterisation of Volatile Organic Compounds Emitted from Indoor Materials and Products Using Small Test Chambers', European Concerted Action, Report 8, EUR 13593 EN, Commission of the European Communities, Luxembourg, 1991.

Figure 2 GC
chromatogram given by
sample of air from a 1 m³
environmental test
chamber containing a
sample of vinyl flooring

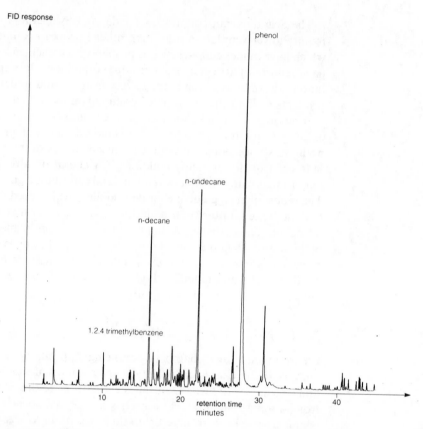

FID response

phenol

n-undecane

n-decane

1.2.4 trimethylbenzene

10   20   30   40

retention time
minutes

over time. Other materials, including other types of vinyl flooring can release a very different mixture of VOCs and these may also have a different emission profile.[34]

International interlaboratory comparisons have been undertaken to assess the comparability of test chamber methods and to increase understanding of the factors that can produce differences between laboratories.[35] Factors such as air speed within the chamber, sample preparation, and analytical procedures have been shown to influence the derived emission rate value.

Currently the European Standards organisation (CEN) is drafting standards for emission testing of building products within committees TC112 and TC264. It is envisaged that these standards will be used to demonstrate compliance with a Requirement of the Construction Products Directive. The relevant Essential Requirement is that of Hygiene, Health, and Environment; and to satisfy the requirement that the product should not emit toxic gas.

Guidance is expected to be issued by the European Commission to define what constitutes a toxic gas. While for a carcinogen there may be no safe level, for

[34] V. Brown, D. Crump, and C. Yu, Proceedings of the International Conference on VOCs, London, October, 1993, p. 283.

[35] M. De Bortoli and A. Colombo, 'Determination of VOCs Emitted from Indoor Materials and Products; interlaboratory comparison of small chamber measurements', European Collaborative Action, Report 13, EUR 15054 EN, Commission of the European Communities, Luxembourg, 1993.

Figure 3 Concentration
of VOCs in a 1 m^3
chamber containing a
sample of vinyl flooring

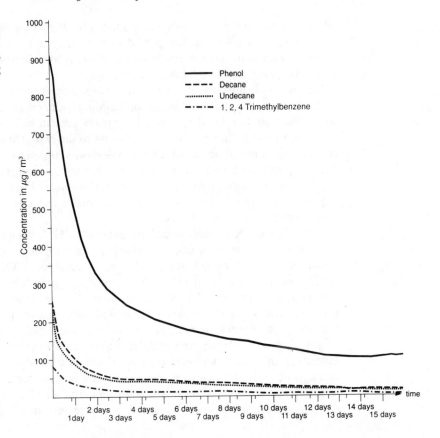

compounds such as irritants a threshold level of effect does occur. In Germany the Hazardous Substance Regulation (1991) specifies a chamber test for the measurement of formaldehyde emission from wood-based products; the steady state concentration in the chamber should not exceed 0.1 ppm.

In several countries, such as the United States and Denmark, voluntary standards have been applied by flooring manufacturers to limit the emission of TVOCs and some individual VOCs including toluene and 4-phenylcyclohexene from products as determined by a chamber test. A proposed scheme for labelling of building products in Denmark, based on the potential of VOC emissions to cause odour and health effects in the indoor environment, is an example of the application of data from chamber tests to the prevention of indoor air quality problems.[36] The principle is to determine the time period required for the emission of VOCs from the product to decline to a rate that results in an acceptable concentration in a standard room. Rates of emission are determined by a chamber test and used to predict concentrations in a standard room (17.4 m^3) with selected environmental conditions (22 °C, 50% RH, air exchange rate 0.5 h^{-1}, air velocity 1–10 cm s^{-1}). The acceptable concentration is based on

[36] P. Wolkoff and P. Nielsen, 'Indoor climate labelling of building materials—chemical emission testing, long-term modelling and indoor relevant odor thresholds', National Institute of Occupational Health, Copenhagen, Denmark, 1993.

odour and mucous irritation thresholds for individual compounds. Products are then ranked according to the period of time they may cause an air quality problem when installed in a new building, allowing identification of low-emitting products which could be labelled accordingly.

Databases of emission rates of organic compounds from building materials are being developed to assist architects and building specifiers to select materials that when used in the building should provide acceptable indoor air quality. Computer models that link the emission rate to the expected resultant pollutant concentration are under development based on ventilation and air movement. More sophisticated models take account of adsorption and re-emission effects of organic compounds in contact with indoor surfaces. Further development is required to improve the models and then make them applicable for use by practitioners such as architects.

It should then be possible to link the expected VOC emission due to materials used in the construction and furnishing of the building with the ventilation conditions at the design stage to predict resulting VOC concentrations. Additional VOC release due to occupant activities, particularly use of consumer products, would need to be incorporated into the assessment and the resulting predicted concentration could be compared with indoor air quality guidelines that define acceptable concentrations of individual pollutants and mixtures. Clearly considerable research and development work remains to fulfil this objective, but already both materials and ventilation standards are incorporating this approach. It is therefore a topic of growing concern for architects, materials scientists, environmental chemists, engineers, and all other disciplines that are concerned with the built environment and environmental health.

Volatile Organic Compounds: The Development of UK Policy

JOHN MURLIS

1 Introduction

The term volatile organic compound (VOC) covers a wide range of light hydrocarbon species with greatly differing environmental impacts. Concern about VOCs first centred on the direct toxicity of specific VOC compounds and policy was sharply focused on protection in the workplace. VOCs are also precursors of photochemical pollution, but these policies, not being aimed at the broad range of VOCs, made little contribution towards solving the problem of photochemical smog. These secondary impacts of VOC emissions have proved more challenging for policy makers, a concern which is reflected in official definitions of the term VOC.

VOC is not a precise term and a number of definitions are in current use. The United Nations Economic Commission for Europe (UNECE), the prime international forum for the development of policy on Transboundary Air Pollution in the European Region, has adopted the following:

> 'for the purposes of the VOC Protocol, Volatile Organic Compounds means all organic compounds of anthropogenic nature other than methane that are capable of producing photochemical oxidants by reactions with nitrogen oxides in the presence of sunlight'.

Similarly, in the UK National Plan for VOC control, these substances are defined as

> 'a large family of carbon containing compounds, which were emitted or evaporated into the atmosphere and can take part in photochemical reactions in the air'.

The California Environment Protection Agency's Air Resources Board defines VOCs as

125

'*hydrocarbon compounds which exist in the ambient air. They contribute to the formation of smog and/or may themselves be toxic. They often have an odour and some examples include gasoline, alcohol and the solvents used in paints*'.

 The common thread in these definitions is a focus on the secondary effects of VOCs. However, direct effects of VOCs remain of concern, with the emphasis shifting from the workplace to the exposure of the public at large to ambient levels of specific VOCs.

The UK has taken an active role in the development of a science-based VOC policy both at national level and internationally. The Department of the Environment's expert working group, the Photochemical Oxidants Review Group (PORG), has produced authoritative work on the mechanisms by which VOCs produce photochemical pollutants and the Department's Critical Loads Advisory Group (CLAG), has developed criteria for assessing the impacts of photochemical pollution on the natural environment.

The 1990 Environmental White Paper declared that 'the Government intends action on air quality to be increasingly based on the definition of acceptable standards for the protection of health and the wider environment', and announced the establishment of an Expert Panel to advise on Air Quality Standards in Britain. The work of the Expert Panel on Air Quality Standards (EPAQS) is focused on health impacts of VOCs and of their secondary products, such as ozone.

In parallel with the development of UK policy the UK has been actively involved in work in international forums to develop an International Policy on VOCs. In the United Nations Economic Commission for Europe (UNECE) an agreement was concluded to a first stage of VOC reductions in the Geneva Protocol of 1991. In the European Union the UK has been an active participant in the development of a harmonized approach to reporting air quality, work which now includes ozone.

2 Effects of VOCs on Man and the Wider Environment

In the past, UK policy on VOCs arose principally out of the concern for their health impacts in the workplace. Regulations therefore focused on the provision of maximum exposure levels for each of the problematic VOCs. More recently, however, concern about the role of VOCs as precursors to a wide range of photochemical pollutants, but in particular ozone, has been a major driver for UK policy, for example on vehicle emissions.

Most recently, the work of EPAQS has highlighted the issue of exposure of the wider public to VOCs in ambient air. Health concerns are focused on specific VOC species. Benzene and 1,3-butadiene are known carcinogens and recommendations for Air Quality Standards for both compounds have been produced by EPAQS.

VOC compounds, as a class, react with NO_x in the presence of sunlight to produce a range of secondary products including ozone and peroxyacetyl nitrate (PAN). Some of these secondary pollutants are known to have health effects. One of them, ozone, has been subject to an EPAQS study because of the effect it has on the human lung, making breathing difficult for sensitive sectors of the population during episodes of high ozone, and because of the synergistic effect it has with other air pollutants. Ozone is an aggressive oxidant which also damages plants and ages surface treatments on building materials. Individual VOC species, however, have greatly different effects on secondary pollutant production. The photochemical reactions which produce, for example, ozone, proceed at a rate which is dependent on the structure of the particular VOC species involved. In order to estimate the impacts of a particular VOC source, information about the mixture of species it produces is essential. Certain VOC species, for example ethylene (ethene), have been found to impact directly on plants by acting as a growth-regulating hormone. Finally, VOCs play a part in global climate change through the contribution they make to background levels of tropospheric ozone.

3 Current VOC Levels in the UK

In order to characterize VOCs in the United Kingdom, the Department of the Environment has taken a twin-track approach. The first has been to produce an inventory of VOC sources and the quantity and speciation of emissions. The second has been to determine ambient levels of VOCs with a monitoring network focused on urban areas. Although inventories of VOCs as a group have been available for the United Kingdom for many years, the need of policy makers to determine feasible UK reduction strategies for UNECE negotiations on the VOC Protocol meant that a more detailed treatment was needed, with a breakdown to individual source categories and species. The production of a fully detailed national inventory of VOC emissions has been a challenging task, since VOCs arise from such a wide range of sources, often as uncontrolled losses, and as a wide range of species and their mixtures. The first task was to identify all VOC sources in the United Kingdom; information on quantities of VOC releases and the species involved continue to be added as data become available. Details of this work are given in the Chapter by Passant (page 51). The major VOC-producing sectors turn out to be transport; the use of solvents in paints, glues, and inks; and the chemical industry. The next challenge for the Emissions Inventory will be to determine the geographic distributions of emissions.

In its monitoring strategy, the Department recognized the need for accurate information on VOC species, anticipating the need for specific VOC Air Quality Standards. The approach taken was to develop a network of instruments capable of resolving the main VOC species of interest at a time resolution matching the time dimensions of the Air Quality Standards, but also sufficient to support work on photochemical pollution. The solution has been to provide on-line gas chromatographic instruments. The design of the network and some preliminary data are given in the Chapter by Dollard and co-workers (page 37).

Certain of these compounds, for example benzene and 1,3-butadiene, are known to arise primarily from motor vehicles. Urban emissions of NO_x are also primarily related to vehicle emissions and the relationships between benzene and 1,3-butadiene and NO_x have been found to be highly consistent. Using this relationship and assumptions about the NO_x/NO_2 ratio, it has been possible to produce maps of the estimated levels of benzene and 1,3-butadiene throughout the United Kingdom from a nationwide survey of NO_2.[1] Generally, levels of

---

[1] J. Stedman, 'Estimated high resolution maps of the United Kingdom air pollution climate', Report AEA/CS/RAMP/16419635/001, AEA Technology, Culham, UK, 1995.

benzene of 1 to 2 ppb as an annual average are predicted, although in some heavily congested urban areas they may rise to 4 or 5 ppb. Eventually it is expected that similar methods will be applied to other VOCs to produce a general picture of the climate of VOC pollution throughout the UK.

Secondary pollutants are monitored in the Department's Ozone Monitoring Network. Results from the network, published by the Photochemical Oxidants Review Group, show the diurnal and annual pattern of ozone pollution in the UK. Typically, high levels are found in summer and daytime hours, with episodes lasting 3 to 4 days. As a broad generalization, episodes arise from the actions of relatively rapidly-acting VOCs, adding to a 'background' ozone level produced by VOCs which act more slowly, in particular methane and incursions from the stratosphere. Results from ozone monitors exposed to air which has not recently passed over heavily populated or industrial areas suggests that the 'background' rises to some 40 ppb in the summer. In order to control levels of photochemical pollution in episodes, however, it is the added ozone which has to be addressed. As a monthly average of hourly averages the excess in the summer amounts to some 20–30 ppb of ozone. At the height of ozone episodes, this can be expected to increase to 50 ppb or more. For much of the year however (and during the night), there is a net depletion of ozone due to anthropogenic NO_x emissions.

The practical consequence of this is that only some 50% of the ozone in episodes is amenable to emission reduction policy, and that reducing episodic ozone will be difficult.

In the UK, then, a clear understanding is developing of current VOC sources and emissions and of the levels of both primary and secondary pollutants they produce in ambient air. It is therefore possible to make a preliminary assessment of the impact of current UK VOC emissions.

4 Effects of Current UK VOC Levels

In order to assess the effects of VOCs at their current levels, clear relationships have had to be established between levels and their impacts. This has been considered by EPAQS who have considered the literature on the health effects of air pollutants and recommended Air Quality Standards. In the case of benzene for example, EPAQS recommended a standard of 5 ppb as a running annual average with 1 ppb as a long term guideline figure to be achieved at a future date. Assessment of current benzene levels, using data from the UK Urban Monitoring Network, suggests that this proposed air quality standard may be exceeded on some occasions near roads which carry a heavy burden of traffic. Levels of benzene are projected to fall further as European vehicle emission legislation reduces VOCs from exhausts.

A similar analysis is possible, both for other primary pollutants, notably 1,3-butadiene, and for ozone. In the case of ozone the air quality standard recommended by EPAQS, 50 ppb as a running 8 hour average, is an ambitious target which is exceeded on at least one occasion in the year at each of the UK monitoring sites. For ozone impacts on the wider environment, an assessment of levels against Critical Levels set for the protection of sensitive vegetation suggests that critical levels for the protection of the natural environment against ozone are generally exceeded in the United Kingdom.

It seems at present, then, that the greatest challenge for VOC policy is posed by the continuing severe impact of secondary pollutants, notably ozone.

5   UK Policy Response

The UK Policy response to these effects is a policy aimed at reducing ambient levels of VOCs in the urban environment and secondary pollutants in rural parts of the UK. As an immediate response the Government has made arrangements to provide information on levels of key hydrocarbon compounds taken from the urban VOC Monitoring Network, and on ozone from the Ozone Monitoring Network, in Air Quality Bulletins available through broadcast information services. Where levels are expected to be high, advice is provided to the public on measures which might be taken to avoid exposure and prevent excess emissions.

As a longer term strategy, the Government has announced a number of measures for air quality management in its recent White Paper 'Air Quality: Meeting the Challenge'. In this document the Government recognizes the central position of Air Quality Standards in an overall framework for air quality management in the UK. This involves a requirement for local authorities to make assessments of air quality and the establishment of Air Quality Management Areas. The Government has undertaken to follow up this recent paper with a more detailed air quality strategy to be published later this year (1995).

In the still longer term the UK plays a continuing part in negotiations on European emission standards for the major VOC sources, and in particular vehicles. Recent examples of proposed EC legislation were the Consolidated Directive on Passenger Vehicle Emissions, the Solvents Directive, and the Directive on Emissions from the storage and transport of petroleum.

In a wider international context, UK has played a role in negotiations in the UNECE on a reduction of transboundary pollution due to VOCs. The first stage measures agreed in 1991 provide for a 30% reduction, based on 1988 emissions, by the year 1999. According to numerical modelling the benefits will be felt the most in North Western Europe and the least in Southern Europe, although ozone problems here are largely concentrated in individual urban areas. The great weakness of this measure in its current form can be seen in the poor return on investment in emission reductions, which is not matched by an equivalent reduction in ozone levels. This comes about because, although the Protocol refers to the different effects of individual VOC species, no attempt has been made in the current form of its primary obligations to allow for them. The analysis has therefore been made on the broad assumption that the reduction required will be made across the VOC inventory as a whole.

For the future, as measures to reduce VOCs become more costly, it will become increasingly important to develop optimal abatement strategies. These should take into account the geographical location of VOC emissions, the reactivity of the different species involved, and the effects of those abatement strategies focused primarily on ozone on the atmospheric levels and deposition of other pollutants.

J. Murlis

6   UK Policy Analysis

The key feature of UK Policy Analysis is the close interactive relationship between the development of policy proposals and the analysis of their potential impacts, based on a clear scientific understanding of the problems to be addressed.

In the case of VOCs the interaction between science and policy has been particularly close, in the main, because of the complex relationships between particular VOC emissions and subsequent ambient levels of secondary pollutants. Even for those VOC species where there is a direct effect, for example benzene or 1,3-butadiene, there is significant challenge. These pollutants arise in the main as part of the mixture of gases in vehicle exhausts. In order to assess the impacts of new emission control technology on benzene, for example, it is essential to understand how a regulation based on limiting VOCs bears on this component in particular. In the case of benzene the main source is road transport, and regulations through the European Community Vehicle Emission Directives have provided emissions standards which control the combined emission of hydrocarbons and NO_x. A considerable research effort has been applied to the measurement (a) of how much of the standard will be taken up by NO_x and how much by VOCs and (b) of the spread of species in the VOC emissions from those vehicles which are expected to meet the new standards. Once the impact of standards on emissions of target species can be characterized, air dispersion modelling provides an indication of how such standards will impact on ambient air quality.

In the case of secondary air pollutants such as ozone, matters are yet more complex, and in order to analyse the impact of future VOC policy, the Department of the Environment has developed a number of powerful modelling tools incorporating the most current understanding of the chemical processes whereby VOCs individually are transformed to ozone.

One of these, the Harwell Photochemical Trajectory Model,[2] explores the production of ozone in an air mass moving across London from the east and travelling westwards during the course of a day. The model assumes that air passing over London picks up both the NO_x and VOC species. A full chemical scheme describing the photochemical behaviour of each VOC species enables the calculation of ozone levels along the trajectory as they react together with the NO_x in the presence of sunlight. By changing the quantity and speciation of the VOCs emitted into the model it is possible to estimate the effects of different regulatory policies. The current state of the trajectory model contains chemistry for 96 hydrocarbon species and has been generalized to produce a national picture, so that changes in the VOC emission inventory can be calculated through to eventual changes in estimated ambient air quality. Models of this kind enable policy makers to estimate the impacts of changes in VOC mixtures, but in order to estimate the impacts of individual VOCs a different approach has been taken.

A trajectory model of a similar kind[3] has been used to examine the impacts of individual species. An air mass from heavily industrialized areas is followed over the course of 4–6 days, and in a number of successive model runs hydrocarbon

[2]   R. G. Derwent and O. Hov, 'Computer modelling studies of the impact of motor vehicle exhaust emissions on photochemical air pollution formation in the UK', *Environ. Sci. Technol.*, 1980, **14**, 1360.

[3]   R. G. Derwent and M. E. Jenkin, 'Hydrocarbons and the long range transport of ozone and PAN across Europe', *Atmos. Environ.*, 1991, **25A**, 1661.

Table 1 Photochemical ozone creating potential (POCP) values* for 69 VOCs evaluated in a two day trajectory model study across the southern British Isles

Compound	POCP value	Compound	POCP value
methane	3	benzene	13
ethane	2	toluene	41
propane	16	o-xylene	41
n-butane	15	m-xylene	78
isobutane	19	p-xylene	63
n-pentane	9	ethylbenzene	35
isopentane	12	n-propylbenzene	25
n-hexane	10	isopropylbenzene	35
2-methylpentane	19	1,2,3-trimethylbenzene	75
3-methylpentane	11	1,2,4-trimethylbenzene	86
2,2-dimethylbutane	12	1,3,5-trimethylbenzene	74
2,3-dimethylbutane	25	o-ethyltoluene	31
n-heptane	13	m-ethyltoluene	41
2-methylhexane	12	p-ethyltoluene	36
3-methylhexane	11		
n-octane	12	formaldehyde	49
2-methylheptane	11	acetaldehyde	33
n-nonane	10	propionaldehyde	28
2-methyloctane	12	butyraldehyde	17
n-decane	8	isobutyraldehyde	43
2-methylnonane	8	valeraldehyde	0
n-undecane	8	benzaldehyde	−40
n-dodecane	7	acetone	9
		methyl ethyl ketone	17
ethylene	100	methyl isobutyl ketone	26
propylene	75		
1-butene	57	methanol	9
2-butene	82	ethanol	4
2-pentene	65	methyl acetate	2
1-pentene	40	ethyl acetate	11
2-methylbut-1-ene	52	isopropyl acetate	17
3-methylbut-1-ene	60	n-butyl acetate	14
2-methylbut-2-ene	71	isobutyl acetate	21
butylene	62		
acetylene	11	methylene chloride	1
		methyl chloroform	0
		tetrachloroethylene	0
		trichloroethylene	6

*Note: The POCP value is the ozone creating potential of a compound relative to ethylene (ethene), expressed as an index where ethylene = 100.

species are removed one at a time from the inventory. After each run the change in ozone levels is assessed. This gives a measure of the contribution of each species to ozone along the trajectory. Results of this work are shown in Table 1. The amount of ozone created by each individual species is related to the amount created by a reference species, ethylene in this case, and given as an index. This index, the photochemical ozone creating potential or POCP, provides a working estimate for policy makers of the impact of the individual VOC species in question. By combining POCPs it is also possible to provide the potential for ozone creation of a particular mixture of VOC species from a given process. Such information can then be used in targeting a control strategy on those species or sources which create most ozone. Once control costs are available it can be also used in evaluating the cost benefits of control of specific processes, sources, or

Table 2 Sectoral POCPs of the various emission sectors and the percentage by mass of VOCs in each ozone creation class

Sector	Sectoral POCP		Percentage mass in each ozone creation class			
			Importance			
	Canada	UK	More	Less	Least	Unknown
Petrol-engined vehicle exhaust	63	61	76	16	7	1
Diesel vehicle exhaust	60	59	38	19	3	39
Petrol-engined vehicle evaporation	—	51	57	29	2	12
Other transport	63	—	—	—	—	—
Stationary combustion	—	54	34	24	24	18
Solvent usage	42	40	49	26	21	3
Surface coatings	48	51	—	—	—	—
Industrial process emissions	45	32	4	41	0	55
Industrial chemicals	70	63	—	—	—	—
Petroleum refining and distribution	54	45	55	42	1	2
Natural gas leakage	—	19	24	8	66	2
Agriculture	—	40	—	—	100	—
Coal mining	—	0	—	—	100	—
Domestic waste landfill	—	0	—	—	100	—
Dry cleaning	29	—	—	—	—	—
Wood combustion	55	—	—	—	—	—
Slash and burn	58	—	—	—	—	—
Food industry	—	37	—	—	—	—

species and it will be of particular value where control rested on substitution of an aggressive VOC with one creating less ozone.

This work highlights first of all the importance of a clear understanding of atmospheric chemistry and secondly, in policy terms, the importance of taking into consideration the different chemical effects of different VOC species or sources.

Table 2 contains an assessment of POCPs for some important UK and Canadian sources. For the UK the immediate challenge is now to incorporate the POCP concept into National Policy development and secondly to ensure that work done in international forums such as UNECE takes full account of speciation and the potential of targeted reductions.

7 Conclusions

UK policy development concerning VOCs rests on a sound scientific understanding of their impacts, origins, and environmental behaviour. UK monitoring networks provide information on the speciation of VOCs in ambient air and the inventory of emissions is targeted on the production of a matching set of estimates of species emissions from different UK sources. Using this information it is possible to assess current pollution levels against targets, and to develop a policy for emissions which will ensure that in future targets for the protection of human health and the environment are met in full.

The major challenge remaining is to ensure that VOC abatement strategies for transboundary problems such as ozone are integrated in an optimal way with local air quality management strategies.

Subject Index